ENGINEERING CHEMISTRY

VOLUME-1

DR. ANINDITA BASAK

BlueRose ONE.com
Stories Matter
NewDelhi • London

BLUEROSE PUBLISHERS
India | U.K.

BlueRose ONE
Stories Matter
NewDelhi • London

For permissions requests or inquiries regarding this publication, please contact:

BLUEROSE PUBLISHERS
www.BlueRoseONE.com
info@bluerosepublishers.com
+91 8882 898 898
+4407342408967

ISBN: 978-93-6452-187-1

Cover Design: Sadhna
Typesetting: Pooja Sharma

First Edition: March 2025

Contents

Water Treatment

Chapter at a glance

- *Use of water for industrial& domestic purpose,sources of water.*
- *Hardness of water,degree of hardness and its estimation (Hehner &EDTA methods),*
- *Ph value of water.*
- *Softening of hard water: Lime soda method,Permutit,Ion exchange and Calgon method.*

Water is the most important factor of life. History reveals that life began with water. Human civilisation too, flourished wherever water was available in abundance.In fact several cities and civilisation disappeared due to shortage of water.

Water is not only the essential for lives of plants and animals but also essential for various industrial process.Our earth includes all sources of water like springs, river, lakes, oceans and underground water.Earth is known as watery planet as it is the only planet in the entire solar system with abundant source of water .About 97% of earth's total water supply lies in the ocean although it is unsuitable for human consumption due to its salinity.About 2% of earth's water is locked in polar ice caps in frozen state.Remaining 1% of earth's water is available in freshwater flowing in rivers, lakes, streams and groundwater possible for human consumption.

SOURCE OF WATER-

The main source of water can be categorised as rainwater, surface water and underground water .

i) RAIN WATER - Purest form of water is available in vapour state in clouds by (precipitation of) evaporation of water.However during its downward journey through the atmosphere it absorbs considerable amount of industrial gases (like CO_2, NO_2,SO_2,Cl_2,) ,organic and inorganic suspended particles.Moreover supply of rainwater being irregular it cannot be used in a planned manner.

ii) SURFACE WATER- The water source which is directly available on the surface of the earth is called surface water. This source may still be like ponds, lakes, or reservoirs (like dams, barrages etc.) or flowing source like springs, rivers, seas etc.

FLOWING SOURCE- Springs, rivers, seas etc are considered as flowing source. Spring channels and streams while flowing over the land dissolve different minerals from soil like chlorides, sulphates, bicarbonates of sodium, magnesium, calcium and iron. It also dissolves organic and inorganic substances like sand, biomass etc. Rivers are fed by rain and spring. So river water contains considerable amount of unwanted matter. Quantity of impurities of river varies from place to place. However with requisite treatment river water can be used for domestic and industrial purpose

iii) SEA WATER- Sea water is the most impure form of water. Sea is called the sink as it contains all the impurities, pollutants bought by river. Moreover due to continuous evaporation the concentration of impurities increases in sea water, makes it unsuitable for human consumption or industrial use

iv) STILL WATER- Dams and barrages are made over the river to conserve and control the river flow throughout the year. The quality water of the reservoir (dams, barrages) is usually better than river water. The water of the reservoir after proper treatment can be used for domestic and industrial purpose, hydroelectricity is also obtained from the water

v) LAKES AND PONDS - These are also storage tanks but the water here contains more organic impurities.It is because decomposition of plants and animals increase the dissolved impurities.Composition of lakes and ponds also vary from place to place.

vi) UNDERGROUND WATER - The water that percolates through the ground forms underground water storage, which is found in form of springs, wells, tubewell etc. As the water passes through different layers of water to underground aquifer, this water is pure but contains dissolve minerals and salts. Chances of bacterial contamination of underground water is minimum.

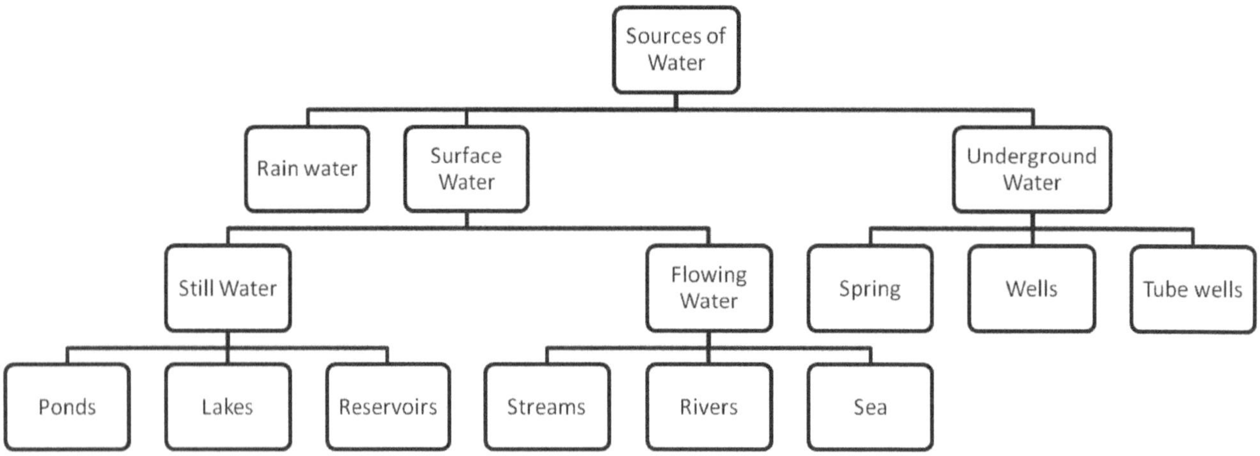

Fig: Different sources of Water.

Water is a vital component of all living flora and fauna on earth.Water is used for domestic purpose like drinking ,cooking, washing, irrigation and industrial use too.The specification required for each use is different ,

for example:

i) POWER GENERATION- Water is used in power generation as steam in turbines and as coolant also.Water used for this purpose should not be hard water, otherwise it will lead to scale formation in boilers, which prevent heat transfer.Untreated water lead to corrosion of boiler material.

ii) TEXTILE INDUSTRY- The water quality required in textile industry is soft water.Water containing dissolve organic matter is undesirable as it produces foul smell.Soft water facilities uniform dyeing.

iii) SUGAR INDUSTRY- In sugar industry also soft water is desired.Hard water produces deliquescent sugar.

iv) BEVERAGE INDUSTRY- Water for this industry should not be alkaline as it neutralises the fruit acids and deteoriates the taste.

v) DOMESTIC USE- In domestic use for cooking soft water is prefered ,the dissolved salts in hard water increases its boiling point.So foods require more time and fuel to be cooked.In washing clothes also soap does not give lather in hard water.

IMPURITIES IN WATER

The impurities present in water may be described as dissolved impurities, colloidal impurities, suspended impurities and presence of micro organisms.

a) **Dissolved Impurities-** During raining water absorbs gases , namely CO_2,SO_2, NO_2 ,H_2S,Chlorides from the atmosphere. while flowing water also absorbs salts like carbonate, bicarbonate, sulphate and chlorides of calcium, magnesium, sodium and iron.The hardness of water is due to these dissolves salts.

b) **Colloidal Impurities-** Water when flow over land surface carries with it clay, finely divided sand particles of size 0.01 - 1μ .Organic matter derived from decomposed plants and animals.This colloidal particles don't settle down.

c) **Suspended Impurities-** In Organic matterials like sand, clay , slit or vegetation between 1-10μm are found in water.This suspended particles add colour, odour and turbidity to water.This suspended particles settle down at the bottom

d) **Micro Organisms-** These include all pathogenic spaces including bacteria, algae, fungi and viruses.This microorganisms are added to waterbody through sewage water mainly.

EFFECTS OF IMPURITIES

The universal solvent nature of water makes it susceptible to contaminations. Water is never found in totally pure form, it will always have some dissolved gases, chemicals, minerals from air soil and rock.These dissolved materials also impart characteristic colour, odour and taste to water.Pure water is generally considered to be flat and tasteless.

The nature of impurities present in water depends on source .For instance presence of minerals like calcium or magnesium cause hardness of water, while iron imparts colour.Again water gets contaminated with presence of arsenic and flouride.We human beings allow run off water containing fertiliser and pesticides to nearby water bodies.The sewage and industrial waste water released in rivers, contaminate it with pathogens.

Thus water from any source needs to be treated prior to its use.The proper water treatment depends upon the purpose on which the water to be used.If it is used for drinking purpose, the water to be treated for removal of objectionable colour, odour, taste and pathogens ,on the contrary water for industrial use must be free from dissolved salts Ca,Mg, Fe and Mn salts and alkalies for use in steam generation, laundries and breweries respectively.

Thus available water is first subject to analysis and then treated suitably.The available water is analysed for parameters like:

i) Hardness

ii) Alkalinity

iii) Dissolved oxygen

iv) Dissolve carbon dioxide

v) Free Chlorine

vi) Chlorides

vii) Silica

viii) Total solids

i) **Hardness of Water -**

Hardness of a water sample is measure of its capacity to precipitate soap i.e; prevent lathering of soap.Water which does not form lather with soap is known as hard water while water which forms copius lather with soap is called soft water.The hardness of the soap is due to presence of magnesium and calcium salts dissolved in it.Soap is sodium or potassium salts of higher fatty acids like oleic ,palmetic,stearicetc dissolves in soft water producing rich lather responsible for cleansing action of soap.In presence of calcium or magnesium salt(as in hard water) they react with soap to scum (precipitate) of the insoluble soap of calcium or magnesium as follows.

$$2C_{17}H_{35}COONa + CaCl_2 \longrightarrow (C_{17}H_{35}COO)_2Ca \downarrow + 2\,NaCl$$

Calcium stearate insoluble soap

$$2C_{17}H_{35}COONa + MgSO_4 \longrightarrow (C_{17}H_{35}COO)_2Mg + Na_2SO_4$$

Magnesium stearate insoluble salt

The soap will lather only after removal of calcium of magnesium ions present in hard water in precipitated as insoluble soap.

Presence of other polyvalent ions such as Al, Fe, Mn etc also cause hardness of water, but the effect is not so pronounce as these polyvalent ions are usually present in complex form not as free ions.

Types of Hardness

Water hardness may be categorised as:

a)Temporary hardness or carbonated hardness

b)Permanent hardness or non-carbonated hardness

a)Temporary hardness or carbonated hardness

Temporary hardness of water is due to presence of calcium or magnesium bicarbonates.This is also known as alkaline hardness(Temporary hardness) which can be removed by boiling ,while bicarbonates decomposed to insoluble carbonates or hydroxides.These precipitates is removed by filtration.

$$Ca(HCO_3)_2 \xrightarrow{\Delta} CaCO_3 + H_2O + CO_2$$

$$Mg(HCO_3)_2 \xrightarrow{\Delta} MgCO_3 + H_2O + CO_2$$

$$MgCO_3 \xrightarrow{\Delta} Mg(OH)_2 + CO_2$$

Addition of equivalent amount of lime $Ca(OH)_2$ may also be used for removal of temporary hardness.

$$Ca(HCO_3)_2 + Ca(OH)_2 \longrightarrow 2CaCO_3\downarrow + 2H_2O$$

$$Mg(HCO_3)_2 + Ca(OH)_2 \longrightarrow MgCO_3\downarrow + CaCO_3\downarrow + 2H_2O$$

b)Permanent hardness or non-carbonated hardness

Permanent hardness of water is due to presence of chlorides and sulphates of calcium,magnesium, iron , aluminium and other heavy metals.Unlike temporary hardness, permanent hardness can not be removed by boiling.Processes like lime soda, zeolite and ion exchange processes are utilised for removing permanent hardness.

Equivalents of Calcium Carbonate(Degree of hardness)

The hardness of water is due to presence of of salts like bicarbonate, chlorides and sulphates of calcium and magnesium.In order to conduct a comparative study of different hardwater samples(degree of hardness) a standardised method is essential.For the purpose $CaCO_3$ is chosen as the standard for expressing hardness of water as equivalent of $CaCO_3$ as because;

i)Molecular mass of $CaCO_3$ is 100,which makes mathematical calculation easier.

ii)$CaCO_3$ is highly insoluble in water hence can be easily precipitated in water treatment processes.

For example let a given hard water sample contains $CaCl_2$ as a cause for its hardness.The soap consuming capacity of this hard water is compared with a standard hard water counting $CaCO_3$i.e;

$$CaCl_2 + 2C_{17}H_{35}COONa \longrightarrow (C_{17}H_{35}COO)_2\,Ca$$

molecular wt.111) sodium sterarte (soap)

$$CaCO_3 + 2C_{17}H_{35}COONa \longrightarrow (C_{17}H_{35}COO)_2\,Ca$$

(molecular wt.100) sodium sterarte (soap)

Thus,

111 parts by weight of $CaCl_2$ would react with the same amount of soap as 100 parts by weight of $CaCO_3$

or,111gm of $CaCl_2$≡100gm of $CaCO_3$

or,1 gm of $CaCl_2$≡100/ 111gm of $CaCO_3$

or, x gm of $CaCl_2 \equiv \frac{100\,x}{111}$ gm of $CaCO_3$

or, x gm of $CaCl_2 \equiv \frac{50\,x}{55}$ gm of $CaCO_3$

Thus to convert any hardness causing salts to $CaCO_3$ equivalent, it should be multiplied by 100/111.

Thus equivalent of $CaCO_3 = \frac{\text{(Equivalent weight of } CaCO_3)}{\text{(Equivalent weight of } CaCl_2)} \times$ (x/mg)

Since $CaCl_2$, $CaCO_3$ and most of the hardness causing salts,

the Equivalent weight = $\frac{Molecular\ weight}{2}$.

The above equation can be written as

Equivalent of $CaCO_3 = \frac{mass\ of\ the\ substance}{molecular\ mass\ of\ the\ substance \div 2} \times 100 \div 2$

or, Equivalent of $CaCO_3 = \frac{mass\ of\ the\ substance}{molecular\ mass\ of\ the\ substance} \times 100$

Equivalent of $CaCO_3 = \frac{\text{(mass of hardness producing substance) x 100}}{\text{(Chemical equivalent of hardness producing substance) x 2}}$

The following table list the multiplication factors for some hardness causing substances.

Table		
SALT	**MOLAR MASS**	**MULTIPLICATION FACTOR**
$Ca(HCO_3)$	162	100/162
$Mg(HCO_3)$	146	100/146
$CaSO_4$	136	100/136
$CaCl_2$	111	100/111
$MgSO_4$	120	100/124
$MgCO_3$	84	100/84
$MgCl_2$	95	100/95
$Mg(NO_3)_2$	148	100/148
HCO_3^-	61	100/61
CO_3^{-2}	60	100/60
OH^-	17	100/ 17x2
H^+	1	100/1x2
$Al_2(SO_4)_3$	342	100/114

The multiplication factor for $Al_2(SO_4)_3$ is calculated as follows:

$$\frac{\text{Equivalent weight of } CaCO_3}{\text{Equivalent weight of } Al_2(SO_4)_3} = \frac{100/2}{342/6} = \frac{50}{57} = \frac{100}{114}$$

Similarly for monovalent salts the multiplication factor -

$$= \frac{2 \text{ X Equivalent weight of CaCO3}}{2\ x\ \text{ x Equivalent weight of monovalent substance}}$$

And equivalent weight of monovalent substance = Molar mass

Thus multiplication factor $= \frac{100}{2 \times molar\ mass}$

UNITS OF HARDNESS

1) Parts per million (ppm) - It is the parts of $CaCO_3$ equivalent hardness per 10^6 parts of water.

2) Milligram per litre(mg/l) - It is number of milligram of $CaCO_3$ equivalent hardness present per litre of water.

1mg/l = 1mg of $CaCO_3$ equivalent hardness per litre of water.

But 1L weighs =100gm

$= 1000 \times 1000$ mg.

$1mg/l = 1mg / 10^6 mg = 1ppm$.

3) Clarke's degree - It is the number of grains (1 grain =1 /7000 lb) of $CaCO_3$ equivalent hardness per gallon (10lb) of water.

or,It is the parts of $CaCO_3$ equivalent hardness per 70,000 parts of water.

1° Clarke = 1 grain of $CaCO_3$ equivalent per gallon of water.

1° Cl = 1 part of $CaCO_3$ equivalent hardness per 70,000 parts of water.

4) Degree French(° Fr) - It is the parts of $CaCO_3$ equivalent hardness per 10^5 parts of water.

1°Fr = 1 part of $CaCO_3$ hardness equivalent per 10^5 parts of water

5) Milliequivalent per litre (meq/l) - Number of milliequivalent of hardness present per litre. Thus :

1meq/l = 1meq of $CaCO_3$ per L of water

$=10^{-3} \times 50$ gm of $CaCO_3$ eq per litre

= 50mg of $CaCO_3$ eq per litre

= 50mg/l of $CaCO_3$ eq

= 50ppm.

Inter - relation of Units -

1 ppm = 1mg/l = 0.1°Fr = 0.07° Cl

1° Fr = 10ppm =10mg/l = 0.7°Cl

1° Cl = 14.3 ppm = 14.3 mg/l = 1.43°Fr

1meq/l =50mg/l = 50ppm = 5°Fr = 0.35° Cl

EXAMPLE 1 :

A sample of water contains $Mg(HCO_3)_2$ = 7.3mq/l,$Ca(HCO_3)_2$ = 16.2 mq/l,

$MgCl_2$ = 9.5mq/l , $CaSO_4$ =13.6mq/l. Calculate its temporary and permanent hardness in degree of Clarke.

solution-

Table :Calculation of $CaCO_3$ equivalent

Constituent	Quantity in mg/l	Multiplication factor	$CaCO_3$ equivalent in mg/l
$Mg(HCO_3)_2$	7.3	100/146	7.3 x 100/146 = 5
$Ca(HCO_3)_2$	16.2	100/162	16.2 x 100/162 =10
$MgCl_2$	9.5	100/95	9.5 x 100/95 =10
$CaSo_4$	13.6	100/136	13.6 x 100/136 =10

∴ $CaCO_3$ equivalents for temporary hardness due to $Mg(HCO_3)_2$ and $Ca(HCO_3)_2$

=(5+10)mg/l

= 15mg/l or 15ppm.

Permanent hardness due to $MgCl_2$ and $CaSO_4$

= (10+10) mg/l or 20ppm.

Interconversion of units -

1 ppm =0.07°Cl

∴ 15ppm =(15 x 0.07)°Cl =1.05°Cl

20ppm =(20 x 0.07)°Cl = 1.40°Cl

Thus the temporary and permanent hardness of the given water sample is found 10 be 1.05°Cl and 1.40°Cl respectively.

EXAMPLE 2:

How many grams of $CaSO_4$ dissolved per litre in water will give 92ppm of hardness

solution-

$$\text{Hardness} = (\text{strength of } CaSO_4 \text{ in mg/l}) = \frac{\text{Chemical equivalent of } CaCO_3}{\text{Chemical equivalent of } CaSO_4}$$

$$\text{or, strength of } CaSO_4 \text{ in mg/l} = \text{Hardness} \times \frac{\text{Chemical equivalent of } CaSO_4}{\text{Chemical equivalent of } CaCO_3}$$

= (92 x 68/50)ppm

=125.12 mg/l

=0.125gm/l

∴ 0.125gm of $CaSO_4$ dissolves per litre of water will give 92 ppm of hardness.

EXAMPLE 3:

200ml of water sample exhibits a hardness equivalent to 25 ml of 0.08N $MgSO_4$. Find the hardness in°Fr.

solution -

No. of gm eqv of $MgSO_4$ = Volume of $MgSO_4$ x Normality

=(25 x 0.08)/1000

=2 x 10^{-3}

Weight of $MgSO_4$ in gm present in 200ml of water

=(2 x 10^{-3}) x equivalent weight of $MgSO_4$ x 1000/200

= 10^{-2} x Eqvwt of $MgSO_4$

Again,

$$\text{Hardness} = \frac{(\text{wt in gm of } MgSO_4 \text{ per litre of water}) \times 50}{\text{eqvwt of } MgSO_4}$$

= 50 X 10^{-2} gm/l

=500mg/l

= 500ppm

1ppm = 0.1°Fr

∴ 500ppm = 50°Fr

The hardness is 50°Fr.

DETERMINATION OF HARDNESS

Determination of hardness of water is very important for both domestic and industrial use.Water hardness is a measure of amount of Ca and Mg salts dissolved in water.It is done by the following methods.

i) O.Hehner's Method

ii)Soap tritration method

iii) EDTA method

i)O.Hehner's method -

This method is used for determination of extent of hardness of water for both temporary and permanent hardness

a)Temporary hardness- As temporary hardness can be removed by boiling ,estimation of temporary hardness is done from the difference in alkalinity of water sample before and after boiling.

$Ca(HCO_3)_2 \xrightarrow{\text{boiling}} CaCO_3\downarrow + H_2O + CO_2\uparrow$

$Mg(HCO_3)_2 \xrightarrow{\text{boiling}} Mg(OH)_2\downarrow + 2\,CO_2\uparrow$

Procedure:

To 50ml of water sample taken in a conical flask, 2-3 drops of methyl orange indicator was added and titrated against N/50 HCl.Let V_1 ml in the volume of acid used.Then 100ml of water sample was taken in a pyrex beaker and evaporated to dryness.It was cooled and 50ml of distilled water was added to it.The solution is warmed and stirred to dissolve the soluble portion left.The resultant solution was filtered and the residue was washed with distilled water.Filtrate and washing were collected and the volume was made up to 100ml in a measuring flask with distilled water.

50ml of this solution was titrated against N/50 HCl using methyl orange as indicator.

Let the volume of acid used is V_2ml.

Therefore,

Volume of N/50 HCl used by temporary hardness present in 50ml of water sample

$= (V_1 - V_2)$ ml

∴ Alkalinity due to temporary hardens present in

50ml of water sample $\equiv (V_1 - V_2)$ml of N/50 HCl

∴ Alkalinity due to temporary hardeness in terms of $CaCO_3$ eqv,

$$= \frac{(V_1 - V_2) \times 1 \times 50}{50 \times 50} \text{ gm/l}$$

$$= \frac{(V_1 - V_2)}{50} \text{ gm of } CaCO_3 \text{ eq/l}$$

$$= \frac{(V_1 - V_2)\, 10^6}{50 \times 10^3} \text{ ppm}$$

$$= 20(V_1 - V_2) \text{ ppm.}$$

b) Determination of permanent Hardness- As permanent hardness of water is due to presence of chlorides and sulphates of calcium and magnesium salts.These salts on boiling with sodium carbonate form insoluble carbonates, which can be removed by boiling

i.e; $CaCl_2 + Na_2CO_3 \longrightarrow CaCO_3\downarrow + 2NaCl$

$MgSO_4 + Na_2CO_3 \longrightarrow MgCO_3\downarrow + Na_2SO_4$

Taking advantage of this known sample of hard water is boiled with a known excess of sodium carbonate.After reaction the residual sodium carbonate is determined by titration with a standard acid .The reduction of Na_2CO_3 added is equivalent to permanent hardness of water.

Procedure:-

To 50ml of supplied water sample taken in a beaker 50ml of N/50 Na_2CO_3 was added and boiled.After cooling, the solution was filtered.The residue on the filter paper was washed with distilled water.The filterate and the washing were collected in a conical flask and titrated with N/50 Hal using methyl orange as indicator.Let the volume of the acid used be V ml.

∴ Volume of N/50 Na_2CO_3 for removing permanent hardness in 50ml water sample =(50 - V)ml

∴ Strength of permanent hardness in $CaCO_3$eqv is

$$= \frac{(50 - V) \times 50}{50 \times 50} \text{ gm of } CaCO_3\text{eqv}$$

$$= \frac{(50 - V)}{50} \text{ gm of } CaCO_3\text{eqv},$$

$$\therefore \text{Permanent hardness } = \frac{(50 - V) \times 10^6}{50 \times 10^3} \text{ ppm}$$

$$=20(50 - V)\text{ppm}$$

i) Soap Titration Method -

Hard water does not give lather with soap(sodium or potassium salts of higher fatty acid) readily. Only after all hardness causing metal ions present in the hard water sample are precipitated, further addition of soap gives lather.Thus total hardness of water sample can be determined by titration.The appearance of stable lather persisting even shaking for about 2 mins mark end point.

$$2C_{17}H_{35}COONa + Ca(HCO_3)_2 \longrightarrow \underset{\text{insoluble Cal. sterate}}{(C_{17}H_{35}COO)_2Ca\downarrow} + 2NaHCO_3$$

$$2C_{17}H_{35}COONa + MgCl_2 \longrightarrow (C_{17}H_{35}COO)_2Mg\downarrow + 2NaCl$$

$$2C_{17}H_{35}COONa + CaSO_4 \longrightarrow (C_{17}H_{35}COO)_2Ca\downarrow + Na_2SO_{4s}$$

Procedure:

The procedure for estimation of hardness of water requires following steps.

a)Preparation of standard hard water- A standard head water of hardness 1gm/l or 1000ppm or 1mg.ml is prepared by dissolving 1.0gm of pure dry $CaCO_3$ in minimum volume of dry HCl.The solution is boiled to dryness to expel excess acid and CO_2.Then the residue is dissolved in distilled water and made up to 1lt solution.

b) Preparation of soap solution - The soap solution is prepared by dissolving 100gm of pure dry castile soap in 800ml of alcohol plus 200ml of distilled water.

c)Standardisation of soap solution -The burette is rinsed and filled with soap solution, 50ml of prepared standard hardwater is taken in a narrow mouth glass stoppered 250ml bottle. Soap solution from burette is added drop by drop to this bottle, shaken and allowed to stand for some time. Care is taken to note whether a lather is formed and the time for which(if any) it persists. Titration is continued till lather is formed after vigorous shaking last for two minutes. Later the volume of soap solution used be V_1ml.

d)Lather Factor of Distilled water - 50ml of distilled water is titrated against soap solution, till lather persist for two minutes.

Let the volume used be V ml. This volume is to subtracted from all litre values.

e) Total Hardness - The above titration is repeated ,taking 50ml of water sample. Let the titre value of soap solution be V_2 ml.

f)Permanent Hardness - With 250ml of water sample taken in a beaker is heated till the volume is reduced to 50ml[with the view to decompose all bicarbonates to insoluble $CaCO_3 + Mg(OH)_2$]. Then it was filtered ,the filtrate and washing with distilled water are collected in a 250ml measuring flask, and the volume is made upto 250ml with distilled water. Taking 50ml of this solution titration is done against soap solution taken in a burette.

Let the titre value of the solution be V_1 ml.

Calculations-

50ml of standard hard water $\equiv (V_1 - V)$ml of soap solution

or, 50 x 1 mg of $CaCO_3 \equiv (V_1 - V)$ml of soap solution.

$\therefore$ 1ml of soap solution $\equiv \frac{50}{(V_1-V)}$ mg of $CaCO_3$eqv.

Also 50ml of hard water $\equiv (V_2 - V)$ml of soap solution

$= \frac{(V_2 - V)}{(V_1-V)}$ mg of $CaCO_3$eqv.

$\therefore$ 1lt 1000ml) of hard water $= \frac{(V_2 - V) \times 1000}{(V_1-V)}$ mg of $CaCO_3$eqv.

$\therefore$ Total hardness of water $= \frac{(V_2 - V) \times 1000}{(V_1-V)}$ mg/l

$= 1000\frac{(V_2 - V)}{(V_1-V)}$ ppm

Also 50ml of boiled water $= (V_3 - V)$ ml of soap solution.

$= \frac{(V_3 - V)}{(V_1-V)} \times 50$ mg of $CaCO_3$eqv.

∴ 1 lt (100ml) of boiled water $= \frac{(V_3 - V) \times 1000}{(V_1 - V)}$ mg of $CaCO_3$eqv.

∴Permanent hardness $= \frac{(V_3 - V) \times 1000}{(V_1 - V)}$ mg/l of $CaCO_3$eqv.

$= \frac{1000(V_3 - V)}{(V_1 - V)}$ ppm.s

Hence ,temporary hardness =(Total - permanent) hardness

$= \frac{1000[(V_2 - V) - (V_3 - V)]}{(V_1 - V)}$ ppm

or,Temporaryhardness $= \frac{1000(V2 - V3)\text{ppm}}{V1}$

iii) **EDTA Method -**

This is a complexometric titration of a water sample with EDTA using Erichrome Black T as the indicator.

The ions involved in causing hardness of water are Ca^{+2}(aq) and Mg^{+2}(aq), can be determined by titration with chelating agent, ethylene diamine-tetraaceticacid(EDTA) usually as a disodium salt ($H_2Y_2^{-2}$) of it.

i.e; $Ca^{+2}(aq) + H_2Y_2^{-2} \longrightarrow CaY^{-2}(aq) + 2H^{+}(aq)$

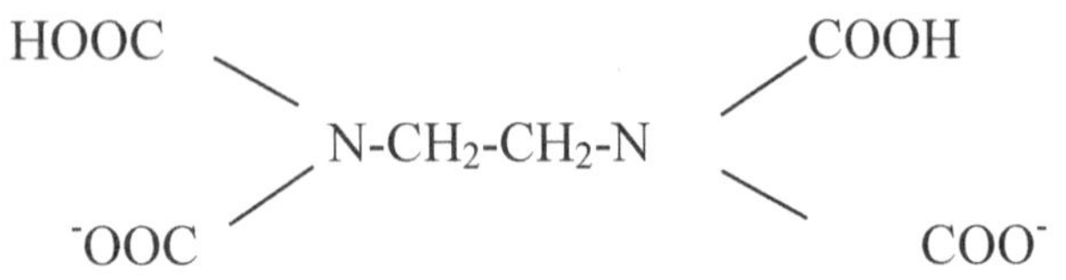

EDTA(The anionic form)

Erichrome Black T is an azodye commonly used as indicator in this titration.Atph 10 ,Ca^{+2}(aq) first forms complex with the indicator as $CaIn^{+}$(aq) which is wine red in colour. As the stronger ligand EDTA is added, the $CaIn^{+}$(aq) complex is replaced by CaY^{-2}(aq)complex which is blue in colour.Thus the end point of titration is indicated by a sharp colour change from wine red to blue.

At equivalent point when the wine red colour changes to blue colour, the reaction is as follows.

[Mg/Ca-Enrichrome Black T] + EDTA ⟶ [Mg/Ca-EDTA] + Eriochrome Black T

(wine red) (blue)

The total amount of EDTA used corresponds to the hardness of water.

Titration using Erichrome Black T as indicator determines total hardness due to Ca^{+2}(aq) and Mg^{+2}(aq) ions.Hardness due to Ca^{+2}(aq) ion is determined by separate titration at a higher ph, by adding NaOH solution to precipitate $Mg(OH)_2$(s) using hydroxy naphthol blue as indicator.

Temporary hardness of water is removed by boiling after removal of precipitate by filtration, the hardness of the filtrate is determined by this EDTA titration corresponds to permanent hardness of water.

Thus: Temporary hardness+Permanent hardness = Total hardness of water.

Experimental Procedure:

i) Preparation of standard hardwater:Dissolve 1mg of pure dry $CaCO_3$ in minimum quantity of dilute HCl and then evaporate the solution on a water bath to total dryness.Dissolve the residue in the distilled water to make 1lt solution.Thus each 1 ml of this solution contains 1mg of $CaCO_3$ eq hardness.

ii) Preparation of EDTA solution : Dissolve 4gm of pure EDTA crystal along with 0.1gm of $MgCl_2$ in 1 lt of distilled water.

iii) Preparation of indicator : Dissolve 0.5gm of Eriochrome black T in 100ml of alcohol.

iv) Preparation of buffer solution : Add 67.5gm of NH_4Cl to 570ml of concentrate ammonia solution and then make it to 1lt of distilled water.

v) Standardisation of EDTA solution : Rinse and fill the burette with EDTA solution. Pipette out 50ml of standard hard water in a conical flask. Add 10-15 ml of buffer solution and 4 to 5 drops of indicator.Titrate with EDTA solution till the wine colour changes to clear blue.Let the volume of EDTA used be V_1 ml.

vi) Titration of unknown hardwater :Titrate 50ml of water sample as above (stepv) .Let the volume of EDTA used be V_2ml.

vii)Titration of permanent hardwater : Take 250ml of water sample in a large beaker.Boil it till the volume is reduced to about 50ml to convert all bicarbonates of Ca^{+2} and Mg^{+2} to hydroxides.Filter and wash the residue with distilled water.Collect the filtrate along with washing and made upto 250ml with distilled water.Then 50ml of this boiled water sample is titrated with EDTA as in step v.Let the volume of EDTA used in V_3ml.

Calculation -

1. Strength of EDTA solution:

1ml of standard hard water(SHW) contains = 1mg of $CaCO_3$

50ml of SHW contains = 50mg of $CaCO_3$

50ml of SHW = V_1 ml of EDTA

V_1ml of EDTA used for = 50mg of $CaCO_3$

1ml of EDTA used for =$50/V_1$ mg/ml of EDTA

Thus strength of EDTA solution $=50/V_1$ mg/ml of EDTA

2. Total Hardness:

Volume of EDTA used for 50ml of sample hard water = V_2 ml

i.e;50ml of sample hard water = 50 x V_2/V_1 mg of $CaCO_3$ eqv

∴ 1lt of given sample water = 50 x V_2/V_1 x 1000/50 mg of $CaCO_3$

$$= \frac{V2}{V1} \times 1000 \text{mg of } CaCO_3$$

∴Total hardness = V_2/V_1 x 1000 ppm

Now 50ml of boiled water = V_3 ml of EDTA

$=V_3/V_1$ x 50 mg of $CaCO_3$ eq.

∴ 1000ml(1lt) of boiled water $=V_3/V_1$ x 1000ppm

∴Permanent hardness $=V_3/V_1$ x 1000ppm

∴Temporary hardness = Total hardness - Permanent hardness

$=1000(V_2/V_1 - V_3/V_1)$ppms

Temporary hardness = $\frac{1000(V_2 - V_1)}{V_1}$ ppm

EDTA method for determination of hardness of water is preferred to other methods due to;

i) Greater Accuracy

ii) Rapid Procedure

iii) Convenience.

Examples:

1.One gm of $CaCO_3$ was dissolved in dil HCl and the solution was diluted to 1 litre with distilled water.100ml of this solution required 56ml of EDTA solution on titration.100ml of the hard water sample required 30ml of EDTA .Again 100ml of the boiled sample water when titrated against EDTA required 15ml EDTA solution. Calculate the temporary and permanent hardness of water.

solution -

Strength of standard hard water =1 mg/ml =1gm/l.

Strength of EDTA-

60ml of EDTA required for = 100ml of standard hard water = 100mg of $CaCO_3$

∴ 1ml of EDTA is required for =100/60mg of $CaCO_3$.

Total hardness-

100ml of sample required = 30ml of EDTA

since 1ml of EDTA is required for 100/60 mg of $CaCO_3$

30ml of EDTA is required for 100/60 x30mg of $CaCO_3$.

So 100ml of sample water contains 50mg of $CaCO_3$

1000ml of sample water contains 500mg of $CaCO_3$

=500mg/lt

Total hardness =500ppm.

Temporary hardness-

After boiling 100ml of sample required 15ml of EDTA.

100ml of sample water contains (15 x 100)/60mg of $CaCO_3$

1000ml of sample water contains 250 mg of $CaCO_3$

Permanent hardness =250mg/l.

∴Temporary hardness = Total hardness - Permanent hardness

= (500 -250)ppm

= 250ppm.

EXAMPLE 2

100ml of sample hard water required 40ml of 0.01M EDTA for titration using Erichrome T as indicator.After boiling 100ml of the same sample, required 20ml of 0.01M EDTA.Calculate total hardness and temporary hardness.

solution -

1 molar solution of EDTA reacts with 1 mole of $CaCO_3$

(As EDTA forms 1:1 complex with metal ions)

100ml of 1M EDTA reacts with = 100gm of $CaCO_3$

1ml of 0.1M EDTA reacts with = 100/1000 x1000/100 = 1 mg of $CaCO_3$.

∴ Strength of EDTA solution = 1mg/ml.

Total hardness-

Since 1ml of EDTA is required for 1 mg of $CaCO_3$

40 ml of EDTA is required for 40 mg of $CaCO_3$

Hence 100ml of sample hard water contains 40mg of $CaCO_3$

1000ml of sample hardwater contains 400mg of $CaCO_3$

∴ Total Hardness = 400mg/l or 400ppm.

Permanent hardness -

20ml of EDTA is required for = 1 x 20mg of $CaCO_3$

Hence 100ml of boiled hard water requires 20mg of $CaCO_3$

1000ml of boiled hard water requires 200 mg of $CaCO_3$

∴ Total permanent hardness is 200mg/l =200ppm.

Temporary hardness -

Temporary hardness = total hardness - Permanent hardens

= (400 - 200)ppm

=100 ppm.

Example 3

Calculate the hardness of a water sample whose 10ml of EDTA solution.20ml of $CaCl_2$ solution whose strength is equivalent 1.5gm of $CaCO_3$ per litre required 30ml of EDTA solution .

solution-

Since 1 litre standard hardwater contains 1.5gm of $CaCO_3$, or, 1ml of standard hardwater contains 1.5mg of $CaCO_3$.

As 30ml of EDTA= 20ml of $CaCl_2$ solution

= 20ml of standard hard water.

=(20x1.5) mg of $CaCO_3$

1ml of EDTA = 30/30 = 1mg of $CaCO_3$

10ml of sample hard water =10ml of EDTA

=10x1 = 10mg of $CaCO_3$eqv hardness

1lt of sample hard water =10ml of EDTA

=10/10x1000 = 1000mg of $CaCO_3$eqv hardness

i.e; Total hardness of water sample is 1000ppm.

P_H Value of Water

The P_H value of water is the deciding factor for its use.In pure water decreasing P_H of about 0.45 occurs with increase in temperature by 25°C, but in the presence of a buffer of bicarbonate-carbonate and hydroxide ions this temperature effect is modified.A direct relation exits between human health and P_H of water to be potable.

ACIDITY :

The low P_Hi.e; acidity of water system is due to dissolved CO_2(which forms carbonic acid) and mineral acidity.Mineral acidity is due to presence of sulphur containing mineral ores(eg:iron pyrites

- FeS_2) as impurities in water.These impurities are oxidised by bacteria(Thiobacillius) in presence of atmospheric oxygen and water to mineral acids, thereby increasing the acidity of water.

$$2S + 3O_2 + 2H_2O \longrightarrow 2\ H_2SO_4$$

$$2FeS_2 + 7O_2 + 2\ H_2O \longrightarrow 2FeSO_4 + 2\ H_2SO_4$$

Acidic water i.e; water with low P_H is not fit for human consumption and also highly toxic for animal and aquatic life too.

ALKALINITY :

Alkalinity of water is measure of its acid neutralising ability. The alkalinity of water due to presence of ions

- Hydroxide only (OH^-)
- Carbonate only (CO_3^{-2})
- Bicarbonates only (HCO_3^-)
- Hydroxides along with carbonates ($OH^- + CO_3^{-2}$)
- Carbonates and bicarbonates ($CO_3^- + HCO_3^-$)

When hydroxides and bicarbonates exist together they react to form:

$$OH^- + HCO_3^- \longrightarrow CO_3^{-2} + H_2O$$

Thus the simultaneous existence of OH^- , HCO_3^- and CO_3^{-2} is ruled out.

The type & extent of alkalinity due to presence of different ions is determined by volumetric titration of the water sample standard sulphuric acid using phenolphthalcein as indicator. Further the titrated water sample is re-titrated against the same standard sulphuric acid with methyl orange as indicator.

i) $OH^- + H^+ \longrightarrow H_2O$

ii) $CO_3^{-2} + H^+ \longrightarrow HCO_3^-$

(i) and (ii): P

iii) $HCO_3^- + H^+ \longrightarrow H_2O + CO_2$

(i), (ii) and (iii): **M**

The volume of acid used upto phenolphthalein end point relates to reactions (i) and (ii) which represents complete neutralisation of OH^- ions and neutralisation of CO_3^{-2} ions to HCO_3^- ions respectively.

And the volume of acid used upto methyl orange end point corresponds to the total reactions representing complete neutralisation of OH^-, $CO_3^{-2}$and HCO_3^- ions.

Procedure -

A known volume of the water sample is titrated with standard acid using phenolphthalein as indicator.At end point when the pink colour completely disappear, methyl orange indicator is added and again titrated with same acid when the yellow colour changes to pink at the end point.The volume of acids consumed is noted.

Observation -

Let the volume of water sample taken =100ml

Volume of acid used up to phenolphthalein end point $=V_1$ ml

Volume of acid used up to methyl orange endpoint $=V_2$ ml

$\therefore$ Total volume of acid used up to methyl orange end point $=(V_1+V_2)$ ml

Normality of the acid used =N/50.

Calculations:

Phenolphthalein alkalinity in terms of $CaCO_3$ equivalents is $N_1V_1 = N_2 \times 100$

$$1/50\ V_1 = N_2 \times 100$$

$$\text{or},\ N_2 = \frac{1 \times V_1}{50 \times 100}$$

Strength in terms of $CaCO_3$ equivalents = N_2 x equivalent of $CaCO_3$ = N_2 x 50 gm/l

$$= V_1/5000(50\text{gm/l}) \times 1000\text{mg/l}$$

$$\text{Phenolphthalein alkalinity } P = \frac{V_1 \times 1000\text{mg/l}}{100}$$

$$= 10\ V_1\ \text{ppm.}$$

2. Methyl orange alkalinity in terms of $CaCO_3$ equivalents

$$N_1(V_1 + V_2) = N_2^1 \times 100$$

$$1/50\ (V_1 + V_2) = N_2^1 \times 100$$

$$\text{or},\ N_2^1 = \frac{1 \times (V_1 + V_2)}{50 \times 100}$$

Strength in terms of $CaCO_3$ equivalents $=N_2^1$ x eq. wt of $CaCO_3$

$$= \frac{(V_1 + V_2)}{5000} \times 50\text{gm/l}$$

or, Methyl orange alkalinity $=M = 10(V_1 + V_2)$ mg/l

$$=10(V_1 + V_2)\text{ppm.}$$

Thus the alkalinity due to presence of different ions can be calculated.

Case 1:

The alkalinity of water is due to presence of OH^- ions only.

$\therefore OH^- + H^+ \rightarrow H_2O$

Hence both phenolphthalein and methyl orange indicator are suitable for end point determination .As irrespective of indicator same volume(V_1ml) of standard acid (say N/50 H_2SO_4) will be needed for OH^- neutralisation in this 1:1 neutralisation reaction.

$\therefore$ M=P = $10V_1$ ppm

$\therefore$ P = M indicates the same water sample is alkaline due to presence of OH^- ions only and the extent is P = M = 10 V_1 ppm.

$\therefore$ P =M indicates the same water sample is alkaline due to presence of OH^- ions only and the extent is P =M = $10V_1$ ppm.

Case 2:

When alkalinity of water is due to presence of CO_3^{-2} ions only.

In this case the alkaline water when titrated against a standard acid (say N/50 H_2SO_4) with phenolphthalein indicator the end point marks completion of reaction A i.e;

$CO_3^{-2} + H^+ \longrightarrow HCO_3^-$ ---------- A.

Let V_1 is the volume of acid consumed .Then on retitration of the titrated liquid with methyl orange indicator will consume same amount of the same standard acid to reach the end point representing neutralisation reaction 'B' as follows;

$HCO_3^- + H^+ \longrightarrow H_2O + CO_2$ - - - - - B.

Let the extra volume of acid consumed is V_2.As reaction A and B are 1:1 neutralisation reactions the volume of acid consumed must be equal.

i.e; $V_1 = V_2$ - - - - - - (1)

As P = 10 V_1 ppm - - - - - — (2)

and M =10 ($V_1 + V_2$) ppm

$\therefore$ M =10 ($V_1 + V_2$) ppm = 20 V_1ppm - - - - (3)

From eq (1),(2) and (3) we can write

$\mathbf{P = \frac{1}{2}M}$ or, $V_1 = V_2$.

Case 3:

The alkalinity of water is due to presence of HCO_3^- ions only.

In this case detection and estimation of HCO_3^- can be done only by with methyl orange indicator and at the end point the completion of neutralisation reaction as follows.

$HCO_3^- + H^+ \longrightarrow H_2O + CO_2$

let the volume of acid consumed $=V_2$ ml.

$\therefore V_1 = 0$ or, $P = 0$.

$M=10\ V_2$ ppm.

$\therefore$ At **P= 0** both OH^- and CO_3^{-2} ions are absent and the alkalinity is due to HCO_3^-only.

Case 4 ;

Alkalinity of water is due to simultaneous presence of OH^- and CO_3^{-2} ions.

Titration of this alkaline water against a standard acid up to phenolphthalein end point(P)marks complete neutralisation of OH^- and neutralization of CO_3^{-2} till HCO_3^- only.

$$\left.\begin{array}{l} OH^- + H^+ \longrightarrow H_2O \\ CO_3^{-2} + H^+ \longrightarrow HCO_3^- \end{array}\right\} P$$

Again titration of the same water sample up to methyl orange end point(M) marks complete neutralisation of OH^- and CO_3^{-2} ions.

$$\left.\begin{array}{l} OH^- + H^+ \longrightarrow H_2O \\ CO_3^{-2} + H^+ \longrightarrow HCO_3^- \\ HCO_3^- + H^+ \longrightarrow H_2O + CO_2 \end{array}\right] M$$

Alkalinity due toCO_3^{-2} = 2(M -P)

Alkalinity due to OH^- = M- 2 (M- P)

= (2P- M)

Case 5 :

Alkalinity of water due to simultaneous presence of CO_3^{-2} ions and HCO_3^-ions.

As explained earlier the acid used till phenolphthalein end point will correspond to neutralisation of CO_3^{-2} till HCO_3^- ions.

And the acid consumed using methyl orange indicator corresponds to

i)neutralisation of HCO_3^- obtained from CO_3^{-2} ions.

ii)HCO_3^-neutralisation which are originally present in the mixture.

Thus alkalinity due to CO_3^{-2} =2P

Alkalinity due to HCO_3^- = (M -2P)

Again as $P = 10V_1$ppm

$M=10(V_1 + V_2)$ppm.

Since $V_1 < V_2$

Thus $P < \frac{1}{2}M$.

The following Table :will summarise the total result for computing the alkalinity due to different ions.

Sl. No.	Result of Titrations	Alkalinity due to OH-	Alkalinity due to CO_3^{-2}	Alkalinity due to HCO_3^-
1	P =0	Nil	nil	M
2	P = M	P or M	nil	nil
3	P = 1/2M(or $V_1 = V_2$)	Nil	2P	nil
4	P>1/2M (or $V_1 > V_2$)	(2P =M)	2(M-P)	nil
5	P<1/2M or ($V_1 < V_2$)	Nil	2P	M-2P

Drawbacks of using high alkaline water -

a)It may lead to caustic embrittlement.

b) It may lead to deposition of precipitates and sludges in boiler tubes and pipes.

Scale and sludge formation in boilers caustic embrittlement ,Primary and Foaming

Solved numerical.

Example 4-

100ml of water sample requires 20ml of N/50 H_2SO_4 for neutralisation to phenolphthalein end point..After this methyl orange indicator was added and further acid required was again 20ml.Calculate the alkalinity of water as $CaCO_3$ in ppm.

i) 100ml of water upto phenolphthalein end point = 20ml of N/50 H_2SO_4.

$$= \frac{20 \ x \ 1}{100 \times 50}$$

$$= \frac{20}{500}N$$

Strength of alkalinity upto phenolphthalein print (in terms of $CaCO_3$ equivalent).

$$P = \frac{20 \times 50}{500} = 0.2gm/l = 200mg/l$$

- 100ml of water upto methyl orange end point =(20ml +20ml of N/50 H_2SO_4).

Normality $N_m = \frac{40}{100} \times \frac{1}{50} = \frac{4}{500}N$

Strength in terms of $CaCO_3$ equivalents

$$M = \frac{4}{500} \times 50 \quad gm/l.gm/l$$

$=0.4gm/l$

$= 400mg/l$.

from (i) and (ii) $P = \frac{1}{2}M$,So alkalinity due to CO_3^{-2} ion only.

$\therefore$ Total alkalinity = 2P =M = 400mg/l

= 400 ppm.

Example 5 -

200ml of water sample required 9.4ml N/50 HCl for complete neutralisation with phenolphthalein indicator. On addition of a few drops of methyl orange to the above water sample and titrated further with the same acid required 21ml of N/50 HCl to reach the end point.

Elucidate the type and extent of alkalinity present in the water.

Solution

200ml of water upto phenolphthalein endpoint

= 9.4ml of N/50 HCl

i.e; 200ml of N_P =9.4ml x N/50

$N_P = \frac{9.4}{200} \times \frac{1}{50}N$

Thus the strength of alkalinity upto phenolphthalein end point in terms of $CaCO_3$ equivalent = P = N_P x 50 x 1000 ppm.

$\Rightarrow$ P = 47 ppm.

200ml of water upto methyl orange end point.

=9.4 + 21 =30.4ml of N/50HCl.

$\therefore$ 200ml x N_M = 30.4ml x N/50

$N_M = \frac{30.4}{200} \times \frac{1}{50} N$

Hence strength of alkalinity upto methyl orange end point in terms of $CaCO_3$ equivalent hardness.

= M = N_M x 50 x 1000 ppm.

M = 152ppm.

Since P < 1/2 M,

=> CO_3^{-2} and HCO_3^- ions are present.

and alkalinity due to CO_3^{-2} =2P = 2 x 47 = 94ppm

alkalinity due to HCO_3^- = M-2P

=152 -94

=58 ppm. (ans)

Example 6:

100ml of a hard water sample neutrilizes 12ml of 0.1(N) HCl with methyl orange indicator .What kind of hardness is indicated .Express the hardness as $CaCO_3$ equivalent.

Solution-

Methyl orange indicator gives the value for temporary hardness present.

Given,Normality of HCl in 0.1N $=N_1$

Volume of HCl in 12ml $=V_1$

Volume of hard water 100ml $= V_2$

Let Normality of hard water $=N_{2s}$

$\therefore N_2=N_1V_1/V_2$

$$= \frac{0.1\times12}{100} N$$

Equivalent weight $CaCO_3 = 50$

$\therefore$ Strength of hardness in $CaCO_3$eqv $= N_2$ x 50gm/l.

$$= \frac{0.1\times12}{100} x\ 50gm/l$$

= 0.6gm/l

Temporary hardness = 0.6gm/l

= 600 mg/l

= 600 ppm

Softening of Hard Water :

Softening of hardwater is the process of complete removal or reduction of hardness causing salts present in water .It is very much essential as hardwater is unsuitable for domestic as well as industrial use .In Industry water is mostly used for cooling or for steam generation process by boiling, hence has to be sufficiently softened before users hard water causes scale and sludge formation, priming and foaming as already discussed. Softening of hardwater can be done either by

i)External treatment or,

ii) Internal treatment.

i) **External Treatment -** In this method ,softening of water is done prior to entry into the boiler. Methods like lime soda. Zeolite or permutit and ion-exchange processes are followed for the purpose.

ii) **Internal Treatment -** In this method water is treated inside the boiler by addition of chemicals. This method is known as sequestration. Sequestration is a process in which an ion is prohibited to exhibit its usual properties, by either converting it to some other form or by forming, complex compound with some other compound.

Internal treatment of water includes:

a)Colloidal conditioning

b)Carbonate conditioning

c)Phosphate conditioning

d)Calgon conditioning.

Colloidal Conditioning-

In low pressure boiler scale formation can be avoided by addition of colloidal substances like kerosene,tanin,glue ,agar-agar to boiler water.These substances get absorbed on the scale forming precipitates and prevent their coalescence and coagulations.These particles remain loose and form non sticky sludges, which can be removed by blow down operations.

Carbonate Conditioning -

Carbonate conditioning process is based on the principle that precipitation of a salt from a solution takes place only when ionic product of the salt exceeds its solubility product.

For example precipitation of a saturated solution of $CaCO_3$ in water can be represented as ,

$CaCO_3 \rightleftharpoons Ca^{+2} + CO_3^{-2}$

(solution.)

Let K_{sp} is the solubility product of the then ionic product is represented as

$[Ca^{+2}][CO_3^{-2}]$. At any given temperature when equilibrium is established.

$K_{sp\ CaCO3} = [Ca^{+2}] \times [CO_3^{-2}]$ - - - - - - - (1)

Under this condition addition of some Na_2CO_3 will increase the $[CO_3^{-2}]$ concentration.

Following Le Chateliar's principle the increase in carbonate concentration is balanced by precipitation of some $CaCO_3$.

Similarly in a saturated solution of $CaSO_4$

$CaSO_4 \rightleftharpoons Ca^{+2} + SO_4^{-2}$

$\therefore K_{spCaSO4} = [Ca^{+2}] \times [SO_4^{-2}]$ - - - - - -(2)

Now in a boiler water saturated with respect to both $CaCO_3$ and $CaSO_4$, then from eq(1) and eq(2) we can say-

$$\frac{K_{sp\ CaCO3}}{K_{spCaSO4}} = \frac{[Ca^{+2}] \times [CO_3^{-2}]}{[Ca^{+2}] \times [SO_4^{-2}]}$$

or $K' = \frac{[CO_3^{-2}]}{[SO_4^{-2}]}$

$\frac{[CO_3^{-2}]}{[SO_4^{-2}]} = K'$

When $[CO_3^{-2}] > K' [SO_4^{-2}]$, then $CaCO_3$ will be precipitated only in preference to $CaSO_4$ as the solubility product of $CaSO_4$can not be attained under these conditions.Thereby deposition of $CaCO_3$ scales is prevented.The sludge which is formed by $CaCO_3$ can be removed by blow drying.

c) Phosphate Conditioning -

In high pressure boilers carbonate conditioning can not be implemented as the CO_3^{-2} ions suffer hydrolysis and produce OH^- ions

$CO_3^{-2} + H_2O \rightleftharpoons HCO_3^- + OH^-$

$HCO_3^- + H_2O \rightarrow CO_2 + H_2O + OH^-$

Thus the concentration of CO_3^{-2} ions decreases followed by probability of saturation of boiler water w.r.t Calcium or magnesium carbonate reduces.Thus precipitation or sludge formation does not takes place.

However the high OH^- concentration may lead to caustic embrittlement.This is avoided by addition of Na_3PO_4, so that calcium is precipitated as the most insoluble compounds of calcium i.e; $Ca_3(PO_4)_2$

$3CaSO_4 + 2\ Na_3PO_4 \rightleftharpoons Ca_3(PO_4)_2 \downarrow + 2\ Na_2SO_4$

Following the same principle as discussed earlier the concentration of $[PO_4^{-3}]$ is maintained at higher level than $[SO_4^{-2}]$ concentration.

Thus

$$\frac{[PO_4^{-3}]^{2/3}}{[SO_4^{-2}]} > \frac{K\ Ca_3(PO_4)_2]^{2/3}}{K_{CaSO4}}$$

or ,$[PO_4^{-3}]^{2/3} = K'[SO_4^{-2}]$

For precipitation of $Ca_3(PO_4)_2$

$[PO_4^{-3}]^{2/3} > K'[SO_4^{-2}]$

The sludge formed ($Ca_3(PO_4)_2$) is removed by blow drying.

It is to be noted that precipitation of calcium takes place in alkaline medium at a P_H value of 9.5 or above.Depending upon the alkalinity of boiler feed water different sodium salts of phosphate are used.This is tabulated as follows.

1.Trisodium phosphate (Na_3PO_4) -Alkaline boiler feed water.

2. Disodium hydrogen phosphate(Na_2HPO_4) - weakly alkaline boiler feed water

3.Sodium dihydrogen phosphate (NaH_2PO_4) - Acidic boiler feed water

Calgon Conditioning -

Calgon the commercial name of sodium hexameba phosphate, when added to boiler water to prevent scale or sludge formation as it forms complexes with scale forming impurity like $CaSO_4$ into soluble complex compounds which are harmless to the boiler.

$$\underset{\text{Calgon}}{Na_2[Na_4(PO_3)_6]} \rightleftharpoons 2Na^+ + [\,Na_2(PO_3)_6]^{-2}$$

$$2CaSO_4 + [Na_4(PO_3)_6]^{-2} \longrightarrow \underset{\text{soluble complex}}{[Ca_2(PO_3)_6]^{-2}} + 2\,Na_2SO_4$$

Only 0.5 to 5ppm of calgon or sodium hexameta phosphate when added to boiler water prevents scale or sludge formation as explained below.

- At low temperature and pressure it forms stable soluble complexes with Ca^{+2} salts.
- At high temperature and pressure it is converted to sodium orthophosphate and combines with Ca^{+2} salts to form loose sludge calcium ortho phosphate, which can be removed by blow down process.

$$2Na_2[Na_4(PO_3)_6] + 6\,H_2O \longrightarrow \underset{\text{sodium orthophosphate}}{6Na_2P_2O_7} + 6\,H_2\uparrow$$

$$CaSO_4 \xrightarrow{Na_2P_2O_7} \underset{\text{Calcium Orthophosphate}}{CaP_2O_7} + Na_2SO_4$$

Calgon conditioning can not prevent the formation of iron oxide or copper deposition.This is achieved by addition of EDTA or its sodium salt (Trylon B) to boiler water .EDTA combines the scale forming salts into soluble complexes.

External Treatment -

External treatment involve softening of raw water prior to feeding in boilers.Principle of this method is to remove hardness causing salts from the raw material as precipitates.The method most commonly used for the purpose are

1)Lime soda process

2)Base- exchange or Zeolite or permutit process

3)Ion exchange or demineralisation process

1)**Lime soda Process-**

This method consists of converting all soluble hardness causing substances to insoluble precipitates by addition of calculated amount of lime ($Ca(OH)_2$) and soda (Na_2CO_3).

When calculated amount of $Ca(OH)_2$ and Na_2CO_3 are added to raw water, it reacts with calcium and magnesium salts to form insoluble precipitates of $CaCO_3$ and $Mg(OH)_2$. These precipitates allowed to settle and are filtered off.

The chemical reactions that takes place during the lime soda treatment are as follows:

-Reactions with lime ($Ca(OH)_2$)-

a)Removal of temporary hardness due to Ca^{+2} and Mg^{+2} bicarbonates

$Ca(HCO_3)_2 + Ca(OH)_2 \longrightarrow 2CaCO_3\downarrow + 2H_2O$

$Mg(HCO_3)_2 + Ca(OH)_2 \longrightarrow MgCO_3 + CaCO_3\downarrow + 2H_2O$

$MgCO_3 + Ca(OH)_2 \longrightarrow Mg(OH)_2\downarrow + CaCO_3\downarrow$

b)Removal of permanent Mg hardness by converting it to permanent Ca hardness

$MgCl_2 + Ca(OH)_2 \longrightarrow Mg(OH)_2\downarrow + CaCl_2$

$MgSO_4 + Ca(OH)_2 \longrightarrow Mg(OH)_2\downarrow + CaSO_4$

c) Removal of dissolved iron and Al salts

$FeSO_4 + Ca(OH)_2 \longrightarrow Fe(OH)_2\downarrow + CaSO_4$

$2Fe(OH)_2 + 2H_2O + 1/2\ O_2 \longrightarrow 2Fe(OH)_3\downarrow$

$Al_2(SO_4)_3 + 3\ Ca(OH)_2 \longrightarrow 2Al(OH)_3\downarrow + 3CaSO_4$

d) Removal of dissolved CO_2 and H_2S

$CO_2 + Ca(OH)_2 \longrightarrow CaCO_3\downarrow + H_2O$

$H_2S + Ca(OH)_2 \longrightarrow CaS\downarrow + H_2O$

e) Removal of free mineral acids

$2HCl + Ca(OH)_2 \longrightarrow CaCl_2 + 2H_2O$

$H_2SO_4 + Ca(OH)_2 \longrightarrow CaSO_4 + 2\ H_2O$

f) Reaction with $NaHCO_3$

$2NaHCO_3 + Ca(OH)_2 \longrightarrow Na_2CO_3 + CaCO_3\downarrow + 2H_2O$

lime or $Ca(OH)_2$ is used to remove hardness (both temporary and permanent) due to Ca^{+2}, Mg^{+2}, Fe^{+2},Al^{+3} salts and mineral acids namely HCl ,H_2SO_4 .

- Reactions of soda (Na_2CO_3) -

$CaCl_2 + Na_2CO_3 \longrightarrow 2NaCl + CaCO_3\downarrow$

$CaSO_4 + Na_2CO_3 \longrightarrow Na_2SO_4 + CaCO_3\downarrow$

Thus the hardness causing soluble salts of calcium and magnesium are precipitated as in soluble calcium salts and removed by filtration. Some times due to fineness of the precipitates , it is required to add coagulants like $Al_2(SO_4)_3$ or $NaAlO_2$ to accelerate sedimentation followed by filtration.

Calculation for requisite quantity of lime and soda .

Depending upon the quantity of dissolved salts present in raw water, requisite amount of lime ($Ca(OH)_2$) and soda (Na_2CO_3) are calculated following above chemical equations.

i) $Ca(HCO_3)_2 + Ca(OH)_2 \longrightarrow 2CaCO_3 + 2H_2O$

From the above reactions it is indicated that one equivalent of $Ca(HCO_3)_2$ requires one equivalent of lime for removal of temporary hardness due to calcium salt.

=> 1 mole of $Ca(HCO_3)_2$ requires 1 mole of $Ca(OH)_2$

=> 162 parts by the weight of $Ca(HCO_3)_2$ requires 74 parts weight of $Ca(OH)_2$.

≡ 100 parts by weight of $CaCO_3$

∴100parts by weight of $CaCO_3$ equivalent hardness requires

≡ 74 parts by weight of $Ca(OH)_2$

1 part by weight of $CaCO_3$ hardness requires ≡ 74/100 parts by weight of $Ca(OH)_2$

Thus for x gm of $Ca(HCO_3)_2$ present , the $Ca(OH)_2$ required is

= 74/100(Amount of $Ca(HCO_3)_2$ in terms of $CaCO_3$ equivalent.)

$=\frac{74}{100}(x \times \frac{100}{162})$

ii) Similarity for $Mg(HCO_3)_2$ the chemical reaction as follows.

$Mg(HCO_3)_2 + Ca(OH)_2 \longrightarrow MgCO_3\downarrow + CaCO_3\downarrow + 2H_2O$

$MgCO_3 + Ca(OH)_2 \longrightarrow Mg(OH)_2\downarrow + CaCO_3\downarrow$

i.e;

$Mg(HCO_3)_2 + 2\,Ca(OH)_2 \longrightarrow MgCO_3\downarrow + 2CaCO_3\downarrow + 2H_2O$

(2 eqv) (2 eqv)

Thus lime requirement will be $=\frac{74}{100}(2 \times Mg(HCO_3)_2$ in term of $CaCO_3$ equivalents)

iii)Similarly for removal of calcium permanent hardness

$CaSO_4 + Na_2CO_3 \longrightarrow CaCO_3 + 2\,NaSO_4$

$CaCl_2 + Na_2CO_3 \longrightarrow CaCO_3 + 2\,NaCl$

Or alternatively

136gm of $CaSO_4$≡ 111gm go $CaCl_2$≡ 100gm of $CaCO_3$≡ 106gm Na_2CO_3

∴ In terms of $CaCO_3$ equivalents , the soda (Na_2CO_3) requirement will be -

$\frac{100}{106}$ [($CaSO_4$ +$CaCl_2$) in terms of $CaCO_3$eqv hardness]

iv)One equivalent of magnesium permanent hardness requires one equivalent of lime and one equivalent of soda.

$MgCl_2 + Ca(OH)_2 \longrightarrow Mg(OH)_2 \downarrow + CaCl_2$

$MgSO_4 + Ca(OH)_2 \longrightarrow Mg(OH)_2\downarrow + CaSO_4$

$\therefore$ Lime requirement $= \frac{74}{100}$ [$MgSO_4$+ $MgCl_2$ in terms of $CaCO_3$ equivalent]

The $CaCl_2$ and $CaSO_4$ so generated reacts with soda.

$CaCl_2 + Na_2CO_3 \longrightarrow 2NaCl + CaCO_3 \downarrow$

$CaSO_4 + Na_2CO_3 \longrightarrow Na_2SO_4 + CaCO_3\downarrow$

$\therefore$ Soda requirement $= \frac{106}{100}$[$MgSO_4$ + $MgCl_2$ in terms of $CaCO_3$ equivalent]

v) $Al_2(SO_4)_3$ and $FeSO_4$ require equivalent amount lime and soda.

$Al_2(SO_4)_3 + 3\ Ca(OH)_2 \longrightarrow 2Al(OH)_3\downarrow + 3\ CaSO_4$

(6 eqv) (6 eqv)

$FeSO_4 + Ca(OH)_2 \longrightarrow Fe(OH)_2\downarrow + CaSO_4$

Lime requirement $= \frac{74}{100}$ [$Al_2(SO_4)_3$+ $FeSO_4$ in terms of $CaCO_3$ equivalent]

Soda requirement $= \frac{106}{100}$[$Al_2(SO_4)_3$+ $FeSO_3$ in terms of $CaCO_3$ equivalent]

vi)Lime reacts with HCl, H_2SO_4 ,CO_2, H_2S etc and thus their respective equivalents are to be considered as above for calculating lime and soda requirement.

vii) $2NaHCO_3 + Ca(OH)_2 \longrightarrow CaCO_3\downarrow + Na_2CO_3 + 2H_2O$

(2 eqv) (2 eqv) (2 eqv)

For 2 equivalents of HCO_3^- reacts with 2 equivalents of lime to generate 2 equivalents of CO_3^-.

viii)Presence of coagulant $NaAlO_2$ does not require lime or soda but yields NaOH in the product as follows:

$NaAlO_2 + H_2O \longrightarrow NaOH + Al(OH)_3$

1 equivalent of NaOH can be considered to be 1 equivalent of lime and hence the corresponding amount to be subtracted from lime requirement, Hence the total sodalime requirement is as follows.

Total lime $Ca(OH)_2$ requirement

$$= \frac{74}{100}\left[2 \times \text{temp Mg} + \text{Temp Ca} + \text{perm Mg} + (\text{salts of Fe} + \text{Al}) + CO_2 + HCl + H_2SO_4 + HCO_3^- NaAlO_2\right]$$ all in terms of $CaCO_3$ equivalents.

Similarly soda (Na_2CO_3) is consumed by $CaCl_2$ and $CaSO_4$ originally present in raw water and by the equivalent amount of these salts formed by trading the water with $Ca(OH)_2$ having $MgCl_2$, $MgSO_4$, $Al_2(SO_4)_3$, $FeSO_4$,HCl, H_2SO_4 etc.

Thus:

100 parts of $CaCO_3$eqv. hardness requires 106 parts of Na_2CO_3.

Total soda requirement = 106/100 [Perm Ca+ Perm Mg+ salts of Fe+ salts of Al+HCl+$H_2SO_4HCO_3^-$ all in terms of $CaCO_3$ equivalent]

Salient features of Lime - Soda Process ;

i)Temporary hardness due to Ca or Mg salts do not require lime only.

ii)Permanent Ca hardness requires only soda

iii) Permanent Mg hardness requires both lime and soda

iv)Fe,Al salts and free acids also need both lime and soda

v)Substances like NaCl, KCl, Na_2SO_4 , SiO_2 , Fe_2O_3 do not impart any hardness to water hence require neither of soda or lime.

vi) If treated water contains OH^-, CO_3^{-2}, then these are formed from excess of lime ($Ca(OH)_2$) and soda (Na_2CO_3) respectively. So these excess amounts should be added to the calculation.

The chemical reactions in lime soda treatment (as detected above)are slow, also the precipitates formed being very fine produces a super-saturated type solution, which results in sludge formation in pipes and boiler tube .These drawbacks are overcome as follows:

i)Thorough mixing of chemicals and water is carried out

ii)Proper time for completion of reactions are maintained

iii)In hot lime soda process temperature of process is increased.

iv)Proper coagulants like $Al_2(SO_4)_3$ or $NaAlO_2$ are added for suitable sedimentation followed by filtration.

Types of Lime - Soda Process,

Depending upon the operation temperature Lime-soda process can be classified as -

Cold lime Soda Process

Hot lime soda process

Cold lime soda Process -

In this process calculated amount of lime and soda is mixed with water at room temperature.The cold lime soda process is further subdivided into following four categories:

i)Intermittent type - Batchprocess

ii)Conventional type

iii)Catalyst or spiractor type   Continuous process

iv)Sludge blanket type

i) **Intermittent type -Batch Process-** This type of cold lime intermittent softner unit consist of this two tanks. Both the two tanks are provided with inlets for raw water and softener chemicals.These tanks are also provided with a mechanical stirrer each and have two outlets for sludge and the other for transfer of softened water to the filtering unit.

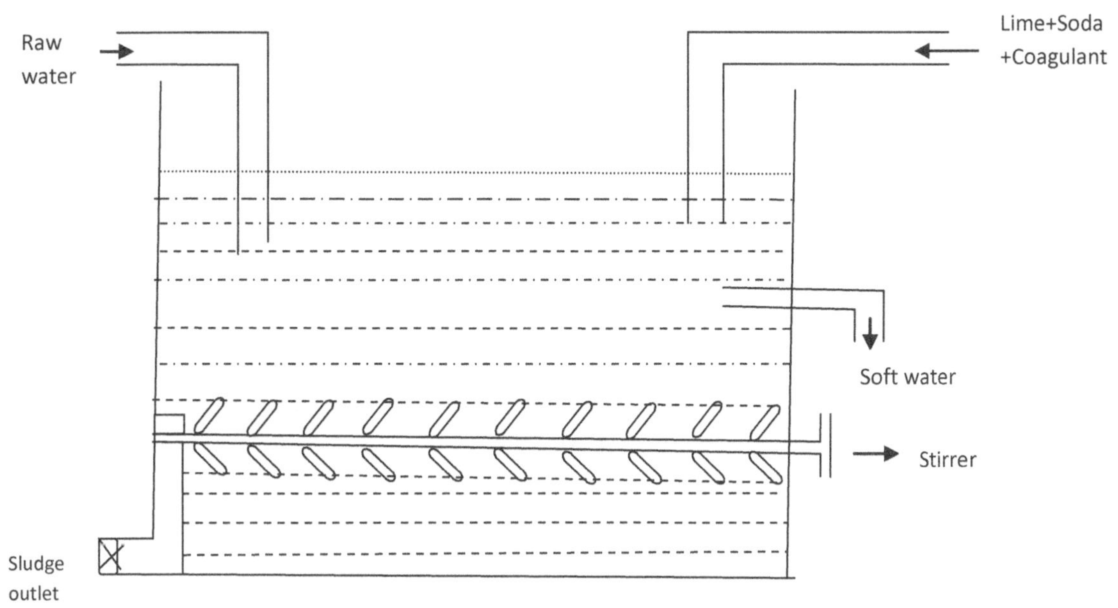

Fig:Batch Process

Raw water and requisite amount of lime or soda and coagulants($NaAlO_2$) are fed into the inlet tank, followed by continuous stirring for through mixing.Some Precipitate of previous treatment are also added during stirring. This hastens fresh precipitate formation. After a fixed time when the precipitation is complete, the stirring is stopped and the precipitate is allowed to settle down. Clear softened water is collected through outlet to the filtering unit while the sludge formed is removed through sludge outlet.

In batch process both the tanks are used alternatively to maintain a continuous flow of softened water. Hence the process is named as intermittent or Batch lime soda process.

ii) **Continuous Softener -** As the name suggests in this method there is a continuous feed of raw water and chemicals (softener and coagulants) to the softener to obtain a continuous flow of softened water.

The continuous softener consists of a big steel tank with two chambers. The inner-chamber is fitted with stirrer and outer chamber with a filtering media of wood fibres.

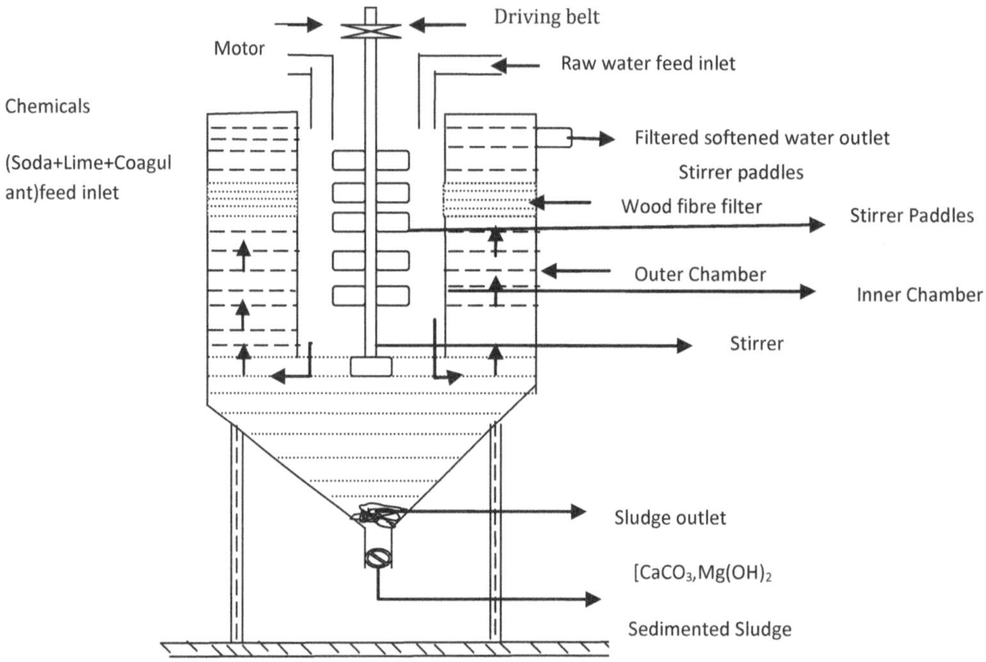

Fig: Continuous cold lime soda softener

Raw water and chemicals namely lime, soda and coagulants is calculated amount are fed into the steel tank through two opposite inlets. The working of stirrers helps in mixing and down flow of the mixture .Precipitation takes place and the sludge settles down at bottom of the outer chamber. Softened water rises up passes through wood filters for ensuring complete removal of sludge. Finally softened water comes out of the outlet and maintains a continuous flow to be used.

$NaAlO_2 + 2H_2O \longrightarrow NaOH + Al(OH)_3\downarrow$

$Al_2(SO_4)_3 + 3Ca(HCO_3)_2 \longrightarrow 2Al(OH)_3 \downarrow +3\ CaSO_4 + 6\ CO_2\uparrow$

iii) **Catalyst or Spiractor type -** In this case a conical tank with two inlets for raw water and chemicals.

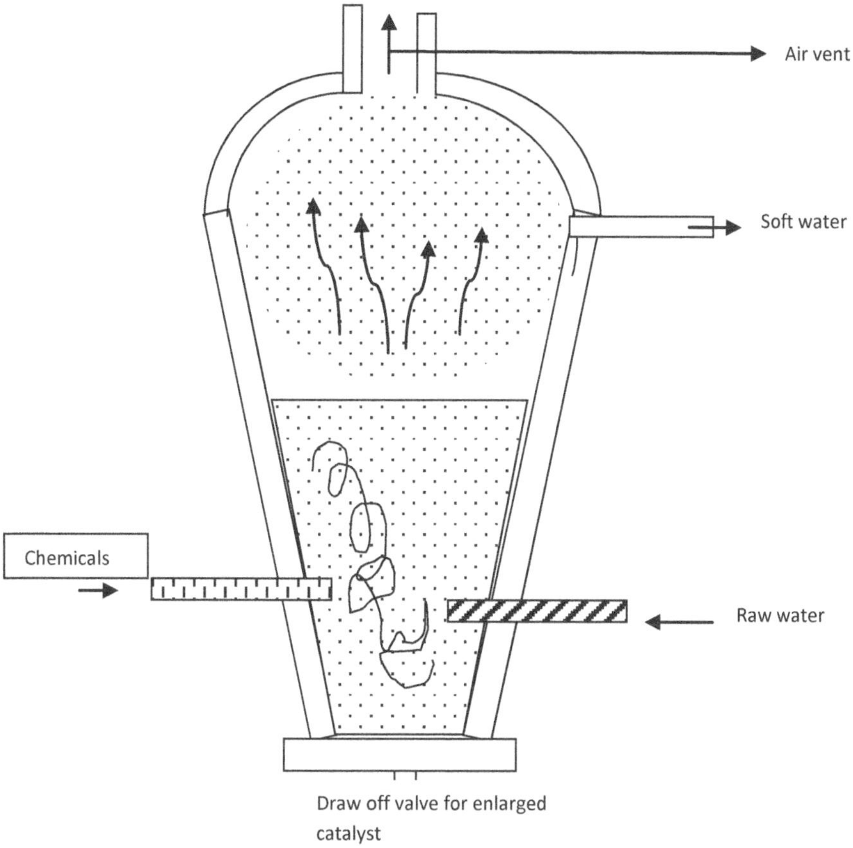

Fig: Spiractor type soda lime process

Raw water and calculated amount of chemicals (lime + soda + coagulants) are fed into the tank through respective inlets. The mixture passes spirally through the catalytic bed. Presence of catalyst speeds up the precipitation and the total process takes approximately 10 ~ 15 minutes. Finally the softened water is collected and sludge is drained off. The catalyst can also be reused.

iv) **Sludge Blanket type -** It is a special type of softener where the raw water is filtered upwardly through the sludge blanket consist of previously formed precipitate in which all the process like mixing, softening and filtration takes place simultaneously. The retention period is 60 to 90 minutes. Silica is well filtered and the unused lime remains in the sludge blanket for reuse.

Hot Lime Soda Process

In this process raw water is treated with chemicals at a elevated temperature of 94 - 100°C .At high temperature the reactions are more effective. The softeners used for hot lime soda process is of two types namely -

- Intermittent type and
- Continuous type.

i) **Intermittent or Batch softener -** This is similar to cold lime soda process. As the reaction is carried out at higher temperature the softner of hot lime soda process, heating coils are installed in it for raising temperature of water.

ii) **Continuous Softner-** The Softener consist of three parts namely reaction tank , conical sedimentation tank and sand filter.

Reaction tank - This part of the softener has three separate inlets one each for raw water , chemicals and super heated steam. After intake ,all three are thoroughly mixed here. Total precipitation reaction gets completed in the reaction tank.

Conical sedimentation tank - From the reaction tank contents are fed into sedimentation tank and the precipitates in the form of sludge settles down here-

a)Sand Filter - It has layers of fine and coarse sand for effective filtration of any residual sludge from softened water.

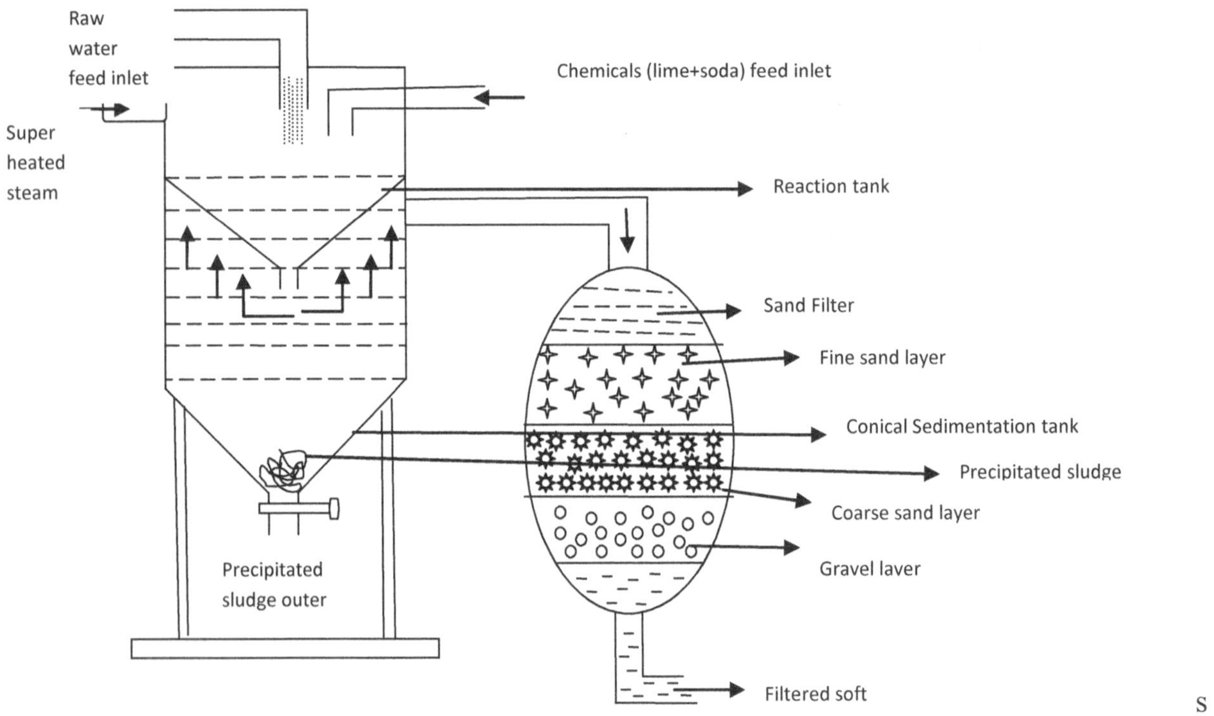

Fig- Hot lime soda process

Advantages of hot lime soda process :

- The reaction time is reduced as rate of softening reaction gets accelerated with increase of temperature close to boiling point of the solution.
- The softening capacity of hot lime soda process increases many fold than the cold process.
- Temporary hardness causing impurities and CO_2 are removed on boiling.
- No coagulant is required as with increase of temperature viscosity of water decreases and sludge and precipitates formed settles down quickly.
- As less chemicals are required the process is economical.

- In this process treated water is alkaline so reduces the corrosion effect and the amount of pathogenic bacteria are also considerably reduced.

However in this process a large amount of sludge is formed and the disposal of this sludge is a problem.

Although depending upon the situation both hot and cold lime soda processe have their own importance. The following table emphasises the main differences between the above two process.

Table Comparison between hot and cold lime process.

HOT LIME SODA PROCESS	COLD LIME SODA PROCESS
Softening is done at elevated temperature 80-150°C	Softening is done at room temperature 25-30°C
Rate of softening reaction is high	It is a slow process
No coagulants are added	Coagulants are needed to facilitate precipitation
Dissolved gases like CO_2 are removed during boiling	Dissolved gases are not removed by simple means
Temporary hardness is also removed on boiling	Lime addition is needed to remove temporary hardness
Viscosity of water reduces on boiling	In no way viscosity decreases
Residual hardness of resultant softened water is 15 - 30ppm	Residual hardness is around 60ppm

SOLVED NUMERICALS

Example -7

Calculate the amount of lime and soda required for softening 20,000 litres of hard water having following salts per litre.

$Ca(HCO_3)_2$ = 9.2mg , $Mg(HCO_3)_2$ = 7.5mg , $CaSO_4$ = 11.2mg

$MgSO_4$ =12.0mg , $MgCl_2$ =2.0mg and NaCl =40mg.

Solution-

From the data given the $CaCO_3$ equivalents are calculated as:

Constituent	Amount mg/l	Multiplication Factor	$CaCO_3$ equivalent mg/l
$Ca(HCO_3)_2$	9.2	100/162	5.68
$Mg(HCO_3)_2$	7.5	100/146	5.14
$CaSO_4$	11.2	100/136	8.23
$MgSO_4$	12.0	100/120	9.99
$MgCl_2$	2.0	100/95	2.10
NaCl	4.0	Neglected	

∴Lime requirement = $\frac{74}{100}$[$Ca(HCO_3)_2$ +2$Mg(HCO_3)_2$ +$MgSO_4$+$MgCl_2$]

=$\frac{74}{100}$[5.68 +2 × 5.14 +9.99 +2.10]mg/l

=20.76 mg/l

Lime required for softening.

20,000 = 20,000 x 20.76

= 415200mg

=0.415kg.

Soda requirement = $\frac{106}{100}$[$CaSO_4$ + $MgSO_4$ + $MgCL_2$]

= $\frac{106}{100}$[8.23 +9.99 +2.10]

=21.54 mg/l

Soda required for softening .

20000 litres = 20,000 x 21.54

= 430784 mg

=0.43kg.

EXAMPLE 8

Calculate the amount of lime and soda required for softening 1 million litres of hard water contains following salts per litre.

$Ca(HCO)_3$ =25ppm ; $Mg(HCO_3)_2$ =144ppm ;$MgCl_2$ = 85ppm ; $CaCl_2$ = 112ppm;

Fe_2O_3 =30ppm ; $Na_2\,SO_4$ =20ppm.

Solution -

From the given data the equivalents of $CaCO_3$ for salts are calculated as.

CONSTITUENT	AMOUNT	MULTIPLICATION FACTOR	$CaCO_3$ equivalent ppm
$Ca(HCO_3)_2$	25	100/162	15.42
$Mg(HCO_3)_2$	144	100/146	98.63
$MgCl_2$	85	100/95	89.47
$CaCl_2$	112	100/80	140

∴Lime requirement =$\frac{74}{100}$[2x$Mg(HCO_3)_2$ +$Ca(HCO_3)_2$ +$MgCl_2$]

[Since Fe_2O_3 and Na_2SO_4 does not contribute to hardness so not considered].

$\therefore$Lime requirement = $\frac{74}{100}$ [2x98.63 +15.42 +89.47]

=223.59 mg/l

Lime required for treatment of 100,000 litre

=100,000 x 223.59

=22.36kg.

As soda is consumed by $CaCl_2$ and $CaSO_4$ originally present and by the equivalent amount of these salts formed on treatment with water and $Ca(OH)_2$

$Ca(HCO_3)_2 + Ca(OH)_2 \longrightarrow 2CaCO_3 + 2\,H_2O$

$Mg(HCO_3)_2 + Ca(OH)_2 \longrightarrow 2\,CaCO_3 + Mg(OH)_2 + 2H_2O$

$MgCl_2 + Ca(OH)_2 \longrightarrow Mg(OH)_2 + CaCl_2$

$CaCl_2 + Na_2CO_3 \longrightarrow CaCO_3 + 2NaCl$

Soda required = 106/ 100 [2 x $CaCl_2$]

=106/100 [2 x 140]

=296.8mg/l

=0.29gm/l

For 1 million litre =1,00,000 x 0.29

=0.029kg

Total Lime and soda requirement for softening of the above hard water are 22.36 kg and 0.029kg respectively.

Example 9-

Calculate the amount of lime (74%pure) and soda (95% pure) required for softening 50,000 litres of hard water contains following salt composition.

$MgCO_3$ =85mg/l ; $CaCO_3$ =40mg/l ; $MgCl_2$ =80mg/l ; $CaCl_2$ =112mg/l ; $Mg(NO_3)_2$=35mg/l ; KCl = 30mg/l

Solution -

From the given data the $CaCO_3$ equivalent is calculated

Constituent	Amount mg/l	Multiplication factor	$CaCo_3$eqv mg/l
$MgCO_3$	85	100/84	101.2
$CaCO_3$	40	100/100	40.0
$MgCl_2$	80	100/95	84.21

$CaCl_2$	112	100/111	100.9
$Mg(NO_3)_2$	35	100/148	23.65
KCl	—	—	—

∴Lime requirement = $\frac{74}{100}$ $[2xMgCO_3 + CaCO_3 + MgCl_2 + Mg(NO_3)_2]$

$$= \frac{74}{100}[2x101.2+40+84.21+23.65]$$

Lime requirement for 50,000 litres =259.2 x 50,000x purity x 50,000 x $\frac{74}{100}$factor

=9.59kg.

Soda Requirement = $\frac{106}{100}[MgCl_2 + MgCl_2 + CaCl_2 + Mg(NO_3)_2]$

$$= \frac{106}{100}[84.21 + 100.9 + 23.65]$$

Soda requirement for 50,000 lts = 221.3 x 50,000 x purity factor

= 221.3 x 50,000x 95/100

= 10.5 kg

Example 10-

Calculate the amount of lime and soda required for softening 50,000 lts of water using 139ppm of $FeSO_4.7H_2O$ as coagulant. The result of analysis of raw and treated water is listed.

Raw water - Ca^{+2} =180ppm , Mg^{+2} =250ppm ;

HCO_3^- =1200ppm ; dissolved CO_2 =20ppm

Treated water - CO_3^- =50ppm , OH^- =48ppm.

Solution -

Determination of $CaCO_3$ equivalent

Constituent	Amount ppm	Multiplication factor	$CaCO_3$ eqv ppm
Raw water			
Ca^{+2}	180	100/40	450
Mg^{+2}	250	100/24	1041.6
HCO_3^-	1200	100/122	983.6

CO_2	20	100/44	45.5
$FeSO4.7H_2O$	139	100/278	50
Treated water			
CO_3^-	50	100/60	83.3
OH^-	48	100/34	141.2

∴Lime requirement = $\frac{74}{100}$ [Mg^{+2}+HCO_3^- + CO_2 +$FeSO4.7H_2O$ +OH^-] x vol of water

$=\frac{74}{100}[1041.6 +983.6 +45.5 +50 + +141.2] \times 50,000$

=83.7kg

Soda requirement -

$=\frac{106}{100}[Ca^{+2} +Mg^{+2} +FeSO4.7H_2O + CO_3^- +OH^- + HCO_3^-]$ x vol of water

$=\frac{106}{100}[\ [450+1041.6 +50+83.3+141.2] \times 50,000$

=93.6kg.

II- BASE EXCHANGE OR ZEOLITE PROCESS OR PERMUTIT PROCESS

A reversible exchange of ions between a solid and liquid phase is known as

ion exchange.It may be exchange of cations only(base exchange) or only anions (anion exchange or ion exchange) between the two different phases solid and liquid.

The principle of the process is as follows. In base exchange a certain class of hydrated alumino silicate mineral acts as solid phase and is capable of exchanging its cations with divalent ions of alkaline earth group or multivalent ions of some of the metals present in water.

In 1976 Cronstedt geologist from Sweden coined the name Zeolite to a naturally occurring hydrated sodium aluminate mineral measuring "boiling stone" as it releases its water of hydration as steam.

Zeolite — $Na_2O.Al_2O_3.xSiO_2.yH_2O$

where x = 2—10 & y = 2 — 6.

Natural Zeolite -

These are green in colour and referred as green sand. The common variety is

Thomsonite — $Na_2O.Al_2O_3.3SiO_3.2H_2O$.

Synthetic Zeolite-

These are prepared by heating together solutions of sodium silicate, aluminium sulphate and sodium aluminate.The most common synthetic zeolite is prepared from feldspar, kaolin, clay and soda. It is white in colour and commonly called as Permutit.

synthetic zeolites are porous with gel like structure and posses higher exchange capacity per unit weight compare to natural zeolite.

The most common synthetic zeolite is prepared from feldspar,kaolin ,clay and soda.It is white in colour and commonly called as Permutit.

Structure:

A zeolite crystal can be considered to be formed from self-linking of several SiO_2 tetrahedral. Each oxygen atom in it is shared with an adjacent one. The empirical formula is thus $(SiO_2)_n$.

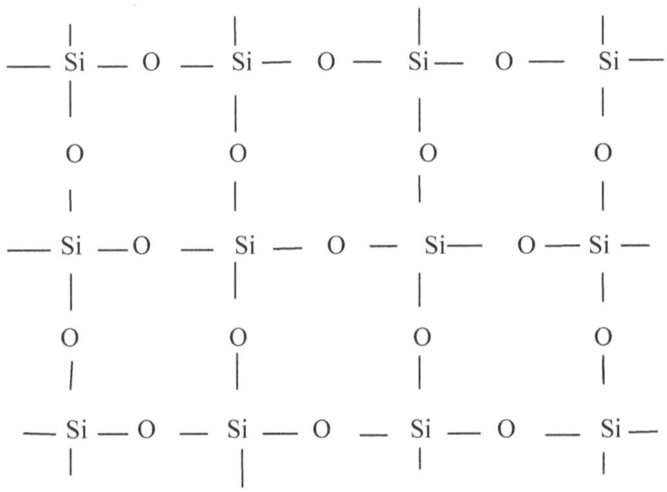

Fig: Zeolite

However some of the Si^{4+} ions are replaced by Al^{3+} ions. In order to balance the charges , extra positive ion such as Na^+, or K^+ must be incorporated for every Al^{3+}introduced.The linking of these tetrahedra results in an open structure with cavities in which Na^+ or K^+ ions are loosely held .It is this porous nature of the structure facilitates free movement of water molecules.

Principle of water softening:

Zeolites can be represented as Na_2Z where Z is the insoluble radical framework. When hardwater is passed through a bed of active granular Na_2Z, the hardness causing ions i.e; Ca^{2+},Mg^{2+}etc, are exchanged for Na^+ ions of zeolite.The extent of exchange increases with increasing valency of the exchanging ions.The reactions taking place during the softening process are as follows:

$Ca(HCO_3)_2 + Na_2Z \longrightarrow CaZ + 2NaHCO_3$

$Mg(HCO_3)_2 + Na_2Z \longrightarrow MgZ + 2NaHCO_3$

$CaSO_4 + Na_2Z \longrightarrow CaZ + Na_2SO_4$

$MgSO_4 + Na_2Z \longrightarrow MgZ + Na_2SO_4$

$CaCl_2 + Na_2Z \longrightarrow CaZ + 2NaCl$

$MgCl_2 + Na_2Z \longrightarrow MgZ + 2NaCl$

Regeneration -

When almost all Na^+ ions are exchanged for Ca^{+2} and Mg^{+2} ions the zeolite bed stop working as it gets exhausted. The exhausted zeolite bed regenerated / reclaimed by following the two steps.

i)Backwashing the zeolite bed with water.

ii)Treating the bed with a concentrated brine (NaCl) solution.

Brine solution is preferred to other solutions like $NaNO_2$,Na_2SO_4,KCl etc as

- Brine is economical
- The products of regeneration reaction (i.e; $CaCl_2$,$MgCl_2$) are highly soluble and easy to rinse out from the zeolite bed.

$CaZ + 2NaCl \rightleftharpoons Na_2Z + CaCl_2$

$MgZ + 2NaCl \rightleftharpoons Na_2Z + MgCl_2$

Exhausted zeolite brine regenerated zeolite washing

Process-

For softening of water by zeolite method ,hard water is percolated at a specific rate through a thick bed of zeolite or permutit packed in a cylindrical steel tank.The hardness causing ions namely Ca^{+2} and Mg^{+2} get exchanged with Na^+ ions and trapped in the zeolite/permutit bed as CaZ and MgZ respectively .The outgoing softened water is rich in sodium salts and collected from the bottom of the tank.

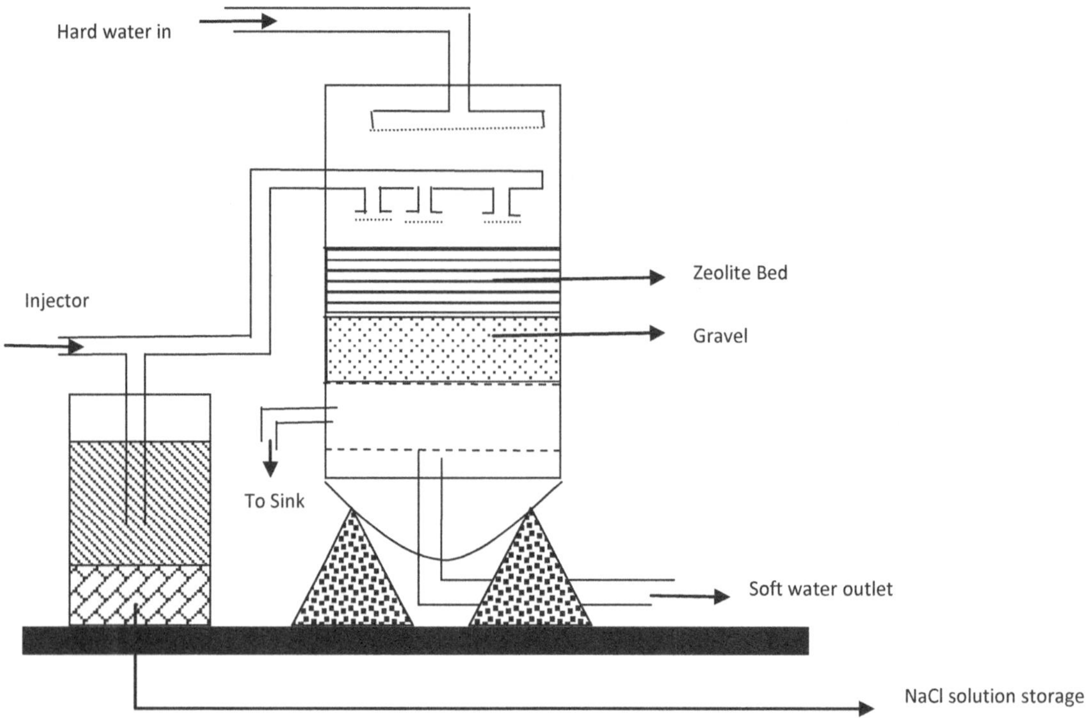

Fig - Zeolite softener

After sometime when the zeolite bed gets completely converted into calcium and magnesium zeolites the flow of softened water ceases.Then the bed is again regenerated as described above and reused. Thus the working of zeolite softners require alternate cycles of softening run and regeneration run.

Limitation of Zeolite softening Process -

- The treated water contains more sodium salts than in lime Soda processes that embrittlement maybe caused.
- It is not suitable for highly turbid waters turbidity and suspended matter clog the pores of zeolite bed thereby decreasing its efficiency and making it inactive.
- The operation P_H of water should be more than 6.5 for efficient softening .Excess acidity and alkalinity may destroy the zeolite bed.
- The method only exchanges Ca^{+2} and Mg^{+2} ions. Presence of the other ions like Fe^{+2} ,Mn^{+2} forms FeZ and MnZ which are difficult of regenerate and bed gets wasted.
- The water to be treated should not be hot as hot water dissolves the zeolite bed
- Anions are not removed by this process.Thus bicarbonate present in hard water gets converted to $NaHCO_3$ and passes with softened water.

$NaHCO_3 \longrightarrow NaOH + CO_2$

If the softened water is used in boiler feed,NaOH causes embrittlement and CO_2 makes water corrosive and acidic. So it is desirable to remove the temporary hardness prior to subjecting to zeolite treatment.

- Compare to ion exchange prices this zeolite process contains 25% more dissolved solids
- Higher cost of the plant and material ate also a major limiting factor.

Advantages -

- The hardness is nearly completely removed water of about 10ppm hardness is produced.
- No sludge is produced in this process.
- The plant is compact and occupies less space.
- It is quite clean and fast process requires less time for softening
- The operation as well as maintenance requires less skill.
- The process automatically adjust itself to water of different hardness.

III. Ion exchange or Demineralisation Process -

The process of complete removal of all ions present in water is termed as demineralization.It is achieved by using ion- exchange resins.

Ion exchange resins are cross linked high molecular weight organic polymers having microporous structure.The functional groups attached to the chain are responsible for ion exchange properties. Depending upon the nature of the groups attached to the polymeric chain the resins may be cationic or anionic ion exchange resins.

Cation exchange Resins(R^-H^+)

These are mainly styrene-divinyl benzene copolymers which on sulphonation or carboxylation becomes capable to exchange their hydrogen ions with cations of water.

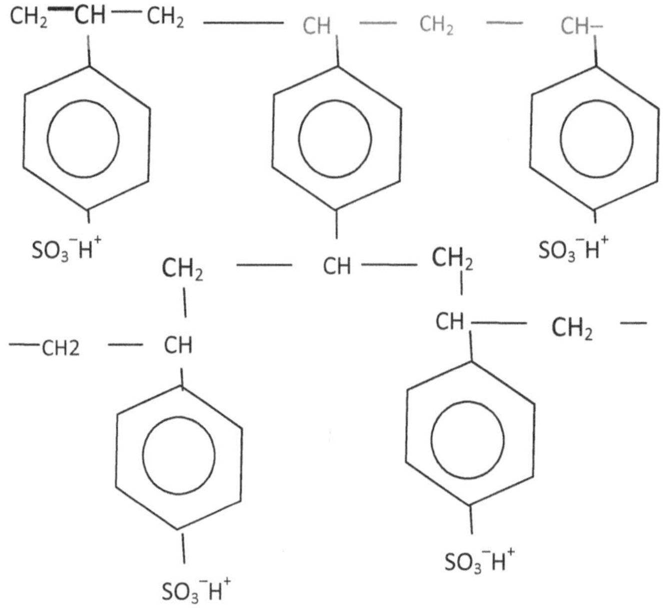

Fig: Sulphonated form of cation - exchanger.

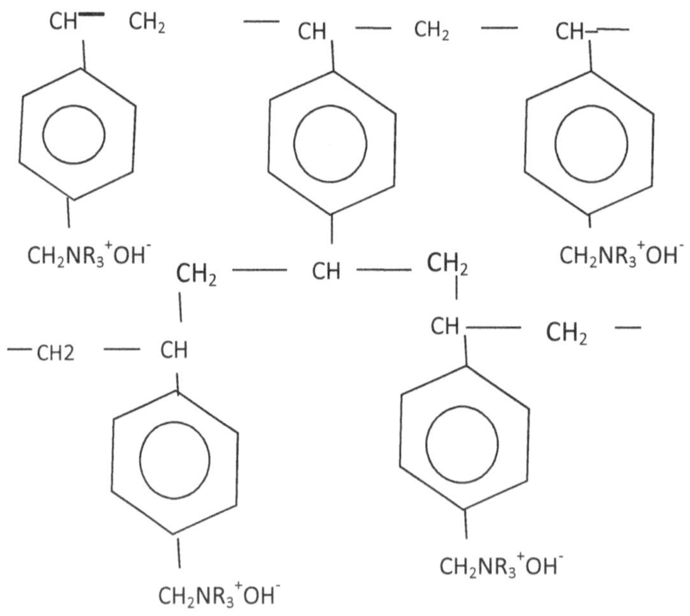

Fig: Anion Exchanger

A number of such resins having acidic functional groups like $-SO_3H$,-COOH or phenolic -OH group have been synthesised, These cation-exchange resins are also known as H-form cation-exchangers.

Aniberlite IR-120 and Dowex -50 are some commercially available cation-exchange resins.

Anion - exchange resins (R^+OH^-)

These are styrene-divinyl benzene or amine-formaldehyde copolymers which contain basic functional groups like amino or quaternary ammonium($-N^+R_3$) or quaternary phosphonium or tertiary sulphonium as an integral part of resin matrix.These copolymer resins after treatment with dil NaOH solution, becomes capable to exchange their OH^- anions with anions of water and known as anion-exchange resins.

Amberlite-400 and Dowex-3 are some commercially available anion exchange resins.

Process-

The ion exchange unit consists of two tanks, one for cation-exchanger and other containing anion exchange resin.The water to be treated is first passed through the cation exchanger, where all the Ca^{+2} and Mg^{+2} cations of water are exchanged with H^+ ions of the resin as follow-

$RH_2 + CaSO_4 \longrightarrow RCa + H_2SO_4$

$RH_2 + MgSO_4 \longrightarrow RMg + H_2SO_4$

$RH_2 + CaCl_2 \longrightarrow RCa + 2HCl$

$RH_2 + MgCl_2 \longrightarrow RMg + 2HCl$

The water treated through cation exchange column is free of Ca^{+2} and Mg^{+2} ions is then allowed to pass through anion exchanger column.Where anions like SO_4^{-2},Cl^-etc present in water are exchanged for OH^-ions of resins.

$R(OH)_2 + H_2SO_4 \longrightarrow RSO_4 + 2H_2O$

$R(OH)_2 + HCl \longrightarrow RCl_2 + 2H_2O$

$R(OH)_2 + H_2CO_3 \longrightarrow RCO_3 + 2H_2O$

Thus the water coming out of the anion exchanger column is free of cations and anions responsible for hardness.Finallly the H^+ and OH^- ions (released from cation exchange and anion exchange columns respectively) get combined to form water.

$H^+ + OH^- \longrightarrow H_2O.$

This water is known as deionised or demineralised water equivalent to distilled water.

Regeneration of Resins :

The exhausted cation exchanger is regenerated as follows -

First the exhausted cation exchanger is back washed and then a 2% H_2SO_4 or HCl solution is passed through it.The regeneration occurs as follow :

$RCa + H_2SO_4 \longrightarrow CaSO_4 + H_2R$

$RMg + H_2SO_4 \longrightarrow CaSO_4 + H_2R$

$RCa + 2HCl \longrightarrow CaCl_2 + H_2R$

$RMg + 2HCl \longrightarrow MgCl_2 + H_2R$

The column is then rinsed with water to remove $CaSO_4$,$MgSO_4$,$CaCl_2$ and $MgCl_2$formed.The washing are rejected.

The exhausted anion exchange resins, it is first backwashed throughly.Then a solution of dilute NaOH is passed through the column.

$RSO_4 + 2NaOH \longrightarrow R(OH)_2 + Na_2SO_4$

Finally the column is rinsed with deionised water and washing are rejected. The regenerated exchanger is now ready for use.

Advantages-

- Water of very low hardness (0-2ppm) is obtained
- Highly acidic and alkaline water can be treated
- As both cation and anions are removed in this process, problems of caustic embrittlement or corrosion are removed.

Demineralised water can be safely used for industrial and domestic as well.

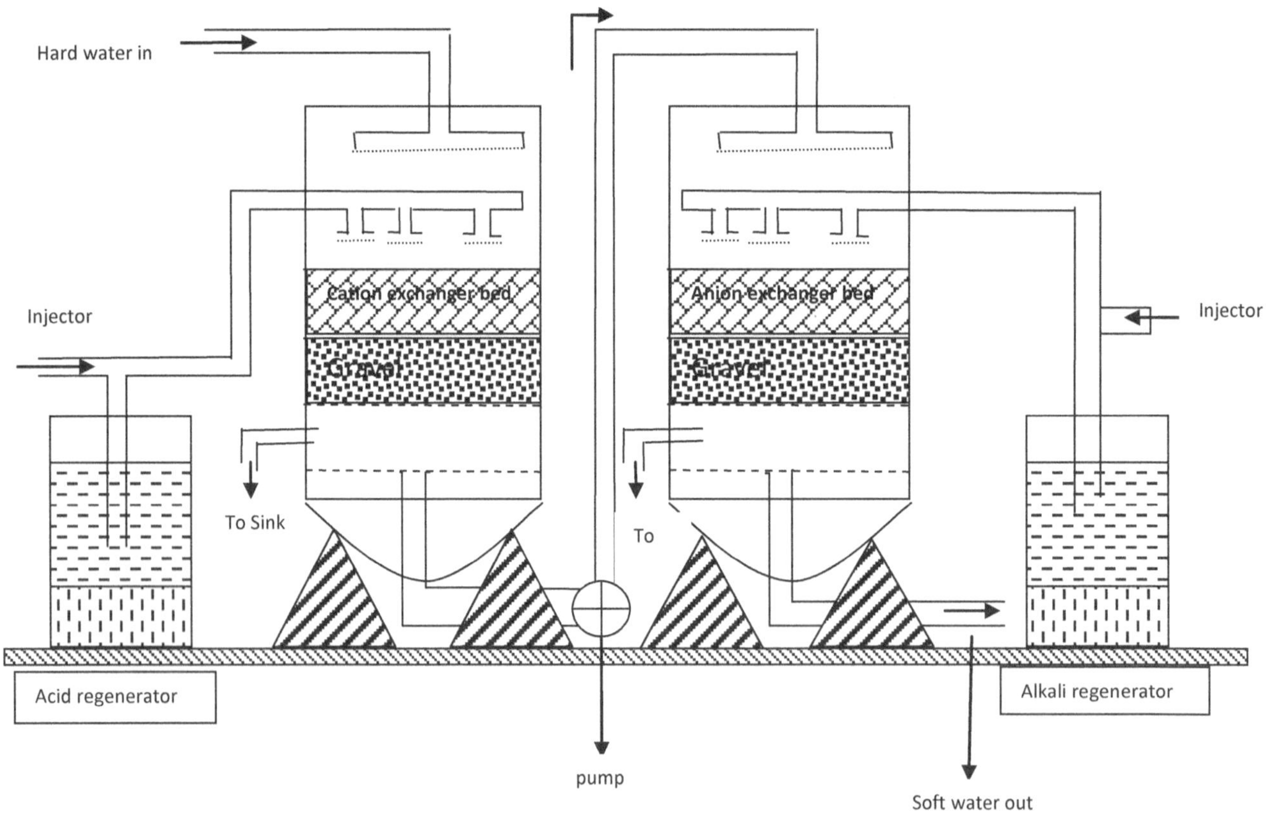

Fig:Softner of demineralised of water.

Disadvantage-

- The equipment and the chemicals required for regeneration are costly
- The efficacy of the system decreases with use of turbid water.

Mixed Bed Deioniser :

In this process the softener consits of a single column filled with an intimate mixture of strongly acidic cation exchanger and strongly basic anion exchanger.On passing hard water through this mixed bed deioniser column it comes in contact with these two types of ion exchangers alternatively several times.The demineralised water coming out of the mixed bed deioniser contains less than 1 ppm dissolved salts.

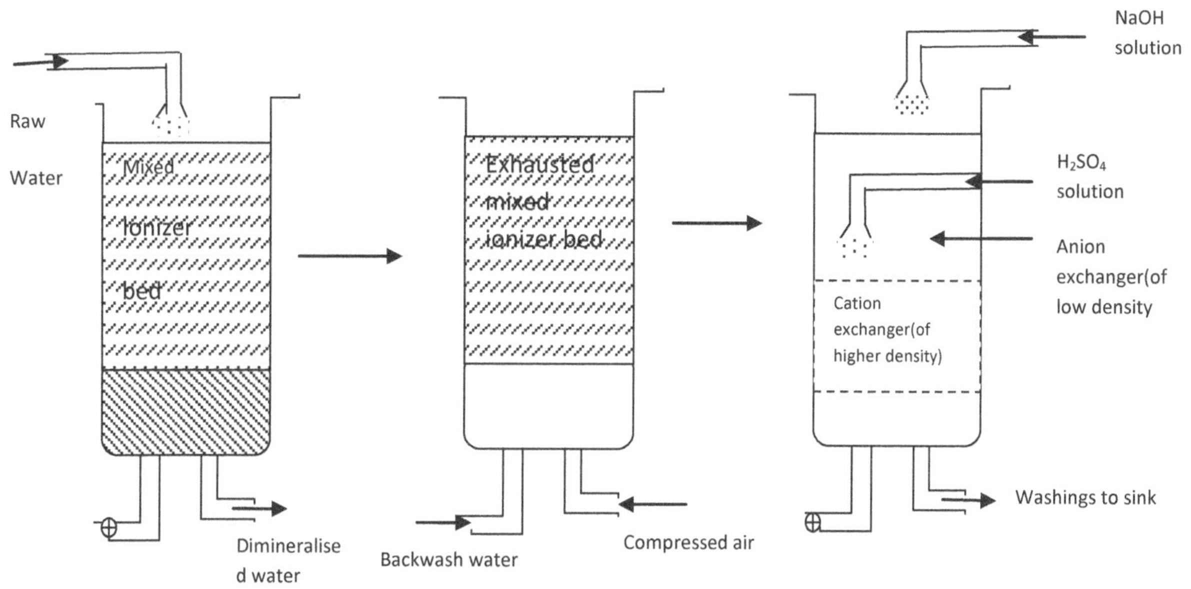

Fig: Regeneration of mixed bed ion exchanger.

Regeneration-

When mixed bed resin get exhausted, the two resins are first separated.Then cationic and anionic resins are backwashed properly and regenerated separately.

The cationic exchanger is treated with strong acid solution like H_2SO_4 while the solution of strong base like NaOH is passed through the anionic exchanger. The two beds are then rinsed with soft water. The washings are rejected. The beds are then mixed again by sending a stream of compressed air from below and become ready for reuse.

Advantages-

- Mixed bed deionisers are more efficient and convenient to use.
- Although it is more costly but it is a simpler and less costly alternative to distillation to be used in laboratories.

- For domestic water softening only cation removal is required hence cationic exchanger is used instead of deioniser.

Table - Comparative study of different water softening process.

Sl. No.	Characteristic	Lime-soda Process	Zeolite Process	Ion-exchanger Process
1	Requirements	Lime,soda and coagulants	Zeolite or Permutit	Cation and Anion exchanger
2	Exchange of ions	No exchange of ions	Exchange of cations takes place	Exchange of both cation and anion takes place
3	Raw water	Prior knowledge of hardness is essential	i)Should be non-turbid ii)Should not be acidic iii)Prior knowledge of hardness is not required	i)Should be non-turbid ii)Both acidic and alkaline water can be treated iii)Prior knowledge of hardness is not required.
4	Automation	Not possible	Possible	Possible
5	Residual Hardness	High (15-50ppm)	Low(0-15ppm)	Least (0-2ppm)
6	Operating Expenses	High	Low	High
7	Capital Cost	Low	High	Very High

Example 11

An exhausted zeolite softener requires 200ltr of NaCl solution containing 200gm/lt NaCl for regeneration. How many ltrs of a sample of water of hardness 400ppm can be softened through it.

solution-

200ltrs of NaCl solution contains = 200x(200gm/l) of NaCl

$$=40000 \text{ x}(\frac{100}{117})\text{eqv of } CaCO_3$$

$$= 34188\text{gm of } CaCO_3\text{eqv.}$$

400ppm of hardness is present in 1 ltr of water,

or,400mg of hardness is present in 1 ltr of water.

$\therefore$ 34188gm = 34188 10^3mg is present in ($\frac{1}{400}$x 34188 x10^3) lt of water

= 85470 lts of hard water.

∴The softener can be soften 85,470 lts of hard water of 400ppm.

Example 12

A zeolite softener was 95% exhausted by removing hardness of 10,000 lts of sample water passed through it. This exhausted zeolite now requires 200ltr of 2% NaCl solution for regeneration. Calculate the hardness of water sample.

solution-

Amount of NaCl contained in 200ltr

$$=(200\text{x } 20)\text{gm}$$

$$= (200 \text{ x}2 \text{ x}\frac{100}{117}) \text{ gm eqv of } CaCO_3$$

$$=3418.8 \text{ gm of } CaCO_3\text{eqv}$$

95% of hardness of 10,000 lt of water.

$$=3418.8 \text{ gm of } CaCO_3\text{eqv}$$

$$1 \text{ lt of water } = \frac{3418.8}{\frac{10000\times95}{100}} \text{gm eqv of } CaCO_3$$

$$=359.87\text{mg eqv of } CaCO_3$$

$$=359.87\text{ppm.}$$

Hence the hardness of water sample is found to be 359.9ppm

BOILER FEED WATER

Industrial use of water mainly comprises of cooling and steam generation purposes.Water used for cooling need not be of high purity. water used for steam generation.ie; boiler feed water requires proper treatment, because for steam generation water is heated in boilers constantly, thus concentration of different unwanted salts/ions also increase progresisvely leading to corrosion with affecting the efficiency of the process.Presence of dissolved impurities in water used as boiler feed is responsible for following different types of problems of boilers.

i)Scale and sludge formation

ii)Boiler corrosion

iii)Caustic embrittlement

iv)Priming and Foaming

v)Carry over

Thus for proper functioning of boilers to produce steam continuously proper treatment of boiler feed water is a necessity..

i) Scale and Sludge Formation -

In a boiler, water is continuously evaporated to form steam.As a result concentration of dissolved salts increases progressively. Finally a stage is reached when ionic product of these salts exceeds their solubility product(beyond saturation point) and the dissolved salts starts precipitating. If the precipitate formed are soft loose and slimy, these are known as sludges, but if the precipitated matter forms a hard adhering crust on the inner walls of the boiler is called scale.

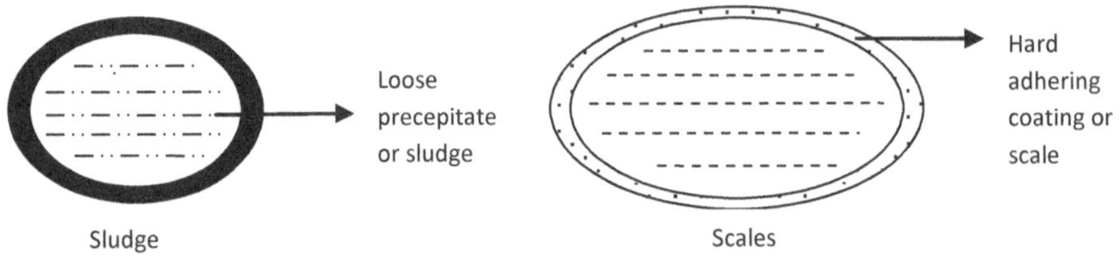

Fig:Sludge and scale in boilers.

Salts like $CaCO_3$,$CaSO_4$, $Mg(OH)_2$, $MgCl_2$ and SiO_2 present in water causes scale and sludge formation.Salts whose solubility in water decreases with increase in temperature(eg-$CaSO_4$) forms scales on hot surfaces and sludges in the cooler regions, but if the solubility of the dissolved salts increase with increase of temperature, then scales are formed on the cooler surface and sludges in hotter regions.

Sludge - Sludges are formed by dissolved salts which have greater solubilities in hot water than cold water e.g; $MgCO_3$,$MgCl_2$,$CaCl_2$,$MgSO_4$etc.Sludge is a soft, loose and slimy precipitate formed within the boiler.

Disadvantages of sludge formation-

i. Wastage of fuel - Sludges being poor conductor of heat, the boilers need to be heated to a higher temperature to get continuous supply of steam, causing excessive use of fuels.

ii. Decrease efficiency -Sludge formation decreases the efficiency of boilers as more heat and time is required to get the desired result thereby sludge formation disturbs the working of boilers.

iii. Chocking of pipes-Sludge formation slowdown water circulation as it settles in the regions such as pipe connection, plug-opening, gauge glass connection etc and finally chocking the pipes.

Prevention of sludge formation -

- Use of soft water can prevent sludge formation.
- Frequent blow down operation(i.e; partial removal of concentrated water with fresh water).The fresh soft water which is added is called make-up water.

Removal of Sludge -

Sludges can be easily removed by scrapping off with a wire brush.

Scale Formation -

Scales are hard deposits of $CaCO_3$,$Mg(OH)_2$ or $MgSiO_3$ or so firmly sticking to the inner walls of the boiler.These scales are so hard that they can not be removed with hammer and chisel.Hence are main source of boiler troubles.

Scales are formed due to -

i)Decomposition of Calcium Carbonate

$Ca(HCO_3)_2 \longrightarrow CaCO_3\downarrow + H_2O + CO_2\uparrow$

$CaCO_3$ scale is soft and scale is the main cause of scale formation in low pressure boiler $CaCO_3$ is soluble in water.

ii)Deposition of $CaSO_4$

The solubility $CaSO_4$ decreases with the increase of temperature.So it is soluble in cold water but insoluble in super heated water.So $CaSO_4$ forms hard scales at hotter regions of the boiler and is difficulty to remove with hammer and chisel.This type scale causes trouble in high pressure boilers.

iii) Hydrolysis of Mg salts

At high temperature of boiler water the dissolved Mg salts get hydrolysed to form soft scales on the inner side of boiler.

$MgCl_2 + 2H_2O \longrightarrow Mg(OH)_2\downarrow + 2HCl\uparrow$

iv) Presence of Silica (SiO_2)

Even if a small quantity of silica is present it causes sticky deposits as $CaSiO_3$ or $MgSiO_3$scales.These scales stick very firmly on the innersidse of boiler surface and difficult to remove.

Disadvantages of Scale Formation

i) Wastage of Fuel - Scales have poor thermal conductivity, so the rate of heat transfer from boiler to inside water is greatly reduced. So for a steady supply of heat to water, excessive or over heating is done thereby increasing fuel consumption. The wastage of fuel is dependent on thickness and nature of the scale formed.

ii) Bagging or Lowering boiler Safety- Due to excessive heating the boiler material becomes softer and weaker and this causes distortion of the boiler material leading to thinning of material.

$3Fe + 4H_2O \longrightarrow Fe_3O_4 + H_2$

(Excesive heating)

The distortion of boiler material due to overheating is called bagging.Bagging makes the boiler unsafe to work with.

iii) Decrease in efficiency - Scales may sometimes deposits in the valves and condensers of the boilers, choke them partially. Thereby decreasing efficiency of boilers.

iv) Danger of explosion -Uneven expansion may lead to cracking of the scales. When superheated water suddenly comes in contact with overheated portion of the boiler body a large amount of steam is formed instantneously, creating high pressure inside the boiler. Development of this sudden high pressure may cause explosion of the boiler specially, when the boiler body is not strong enough due to bagging.

Removal of Scales

- Soft scales are removal with scrapper or wire brush.
- Brittle or hard scales are removed by giving thermal shocks(heating & sudden cooling)
- Loose scales are removed with blow down operation.
- Hard and adherent scales are removed by treatment with chemicals added to water ,for example $CaCO_3$ scales are dissolved in 5-10% HCl. $CaSO_4$ scales are removed by dissolving in EDTA.As Ca-EDTA complex is highly soluble in water.

Prevention of Scale Formation.

The scale formation can be minimised by subjecting the boiler feed water to treatment like external treatment and internal treatment.

Table :Comparative study of Sludge and Scale formation.

Sl.No.	SLUDGES	SCALES
1	Sludges are soft, loose and slim precipitate	Scales are hard deposits
2	Being non-adherent deposits can be easily removed	Scales stick very firmly to the inner surface of boiler hence difficult to remove
3	Formed by dissolved salts like $CaCl_2$,$MgCl_2$,$MgSO_4$,$MgCO_3$ etc.	Formed by dissolved salts $CaSO_4$,$Mg(OH)_2$ etc.
4	Formed at comparatively colder region of the boiler	Generally formed at the hotter portion of the boiler
5	They decrease the boiler efficiency	Decrease boiler efficiency
6	Can be removed by blow down operation	Cannot be removed by blow dry operation

CAUSTIC EMBRITTLEMENT -

Caustic embrittlement is a type of boiler corrosion caused due to use of highly alkaline water in boilers. It makes the boiler material brittle.

During softening of water by lime soda process small amount of free Na_2CO_3 is usually present in softened water. In high pressure boilers this Na_2CO_3 decomposes to give NaOH, which makes the boiler water caustic.

$$Na_2CO_3 + H_2O \longrightarrow 2NaOH + CO_2$$

This caustic water flows into minor hair cracks present in the innersole of boiler by capillary action. On evaporation of water, concentration of caustic soda increases progressively with time. This caustic soda attacks the surrounding area eg; the interganular cracks within the cracks and corroding boiler material or dissolving iron of the boiler as sodium ferrate.

$$3Na_2FeO_2 + 4H_2O \longrightarrow 6NaOH + Fe_3O_4\downarrow + H_2$$

The regeneration of NaOH enhances further dissolution of iron to precipitate as Fe_3O_4.

This makes the boiler body brittle, causing embrittlement of boiler parts mainly stressed one like bends, joints, riverts,etc causing failure of boiler.

Iron at rivers,bends, joints etc	Concentrated NaOH solution	Dilute NaOH solution	Iron at plane surfaces

Caustic embrittlement of iron can be well explained considering the following concentration cell formed.

The iron surrounded by dil NaOH becomes the cathodic side while the iron in contact with concentrated NaOH becomes anodic part, which gets corroded.

Prevention of caustic embrittlement

Caustic embrittlement can be prevented.

i)By using $Na_3(PO)_4$ as softening agent instead of Na_2CO_3 during external treatment of boiler feed water.

ii)By addition of tannin or lignin to boiler water which blocks the hairline cracks in the boiler body and thereby preventing infiltration of caustic soda solution into these areas.

iii)By addition of sodium sulphate to boiler water as it also blocks hair cracks.

It has been observed that caustic cracking can be prevented when Na_2SO_4 is added to boiler water in following ratio $\frac{\text{Na2SO4 conc}}{\text{NaOH conc}}$ is kept as 1:1; 2:1 and 3:1 in boilers working at pressures upto 10,20 and above 20 respectively.

Priming and Foaming

When steam is produced rapidly in boiler, some droplets of liquid water are carried along with the steam. The Phenomenon of wet-steam formation i.e; steam being associated with liquid water droplets is called Priming.

The liquid contamination in steam is expressed in percentage by weight of steam for example if steam contains 0.5% moisture, its steam quality will be reported as 99.5%.

Priming is caused by -

- The high steam velocity.
- Very high water level in boiler.
- Sudden steam demands leading to sudden drop of pressure in the steam line.
- Faulty design of boiler.
- Pressence of considerable quantities of dissolved solids and suspended impurities.

Prevention for Priming -

- Controlling rapid change of steam velocities.
- Proper boiler design.
- Maintaining low water level in boiler.
- Ensuring proper softening.
- Filtration of boiler water carried over to the boiler by blowing off sludges or scale from time to time.

FOAMING -

It can be be defined as the formation of small but persistent foam or bubbles at the water surface of the boilers. Foaming is caused by difference in concentration of solutes or suspended matter between the surface film and the bulk of the liquid. Any material which lowers the surface tension of water(like oil and alkali present in boiler feed water react to form soap) will collect at the interface and increase the foaming tendency of the liquid. Depending upon foaming capability water is classified as foaming water.

Foaming water-The water which produce foam even in two days if blowing off operation is not done.

Semifoaming water - This water does not produce any form in locomotive boilers for two days.

Non foaming water- It is that water which does not produce any foam in locomotive boilers for one week.

Prevention of Foaming

Foaming can be avoided by -

- Addition of antifoaming agents which reduces the surface tension and thereby decreases foaming.Foreg: Castor oil, polyamide antifoamers.
- Removal of foaming agent like oil and grease from boiler water.This is achieved by addition of coagulants like sodium aluminate, ferrous sulphate etc.

Carry Over - Disadvantages of Priming and Foaming

The phenomenon of carrying of water along with impurities by steam is known as carry over. These are mainly due to priming and foaming.Which usually occurs together. Priming and foaming are undesirable because

- Dissolved salts or suspended solids in boiler water are carried by the wet steam to superheaters and turbine blades, where they get deposited as water evaporates. These deposits decreases the efficiency of boiler.
- Dissolved salts may enter into parts of other machinery and decrease their life.
- The maintenance of boiler pressure becomes difficult as foaming produces virtual height of water column..

BOILER CORROSION:

Boiler corrosion is nothing but decay or disintergration of boiler body material due to chemical or electrochemical reaction with its environment. In general when corrosion occurs, the boiler body which is made up of iron combines with water to form $Fe(OH)_2$ that reacts with oxygen to form $Fe(OH)_3$ or $Fe(OH)_2$ that reacts with oxygen to form $Fe(OH)_3$ or rust.

$Fe^{+2} + 2H_2O \longrightarrow Fe(OH)_2 + H_2$

$4Fe(OH)_2 + O_2 \longrightarrow 2[Fe_2O_3.2H_2O]$(rust)

The main reasons for boiler corrosion are due to presence of

i)Dissolved oxygen

ii)Dissolved carbon dioxide

iii)Mineral acids.

i) Presence of dissolved oxygen

Water carry about 8 ml of dissolved oxygen per litre at room temperature. At the prevailing high temperature the dissolved oxygen attacks the boiler body to form ferric oxide (rust)`as per above chemical equation.

Removal of dissolved oxygen- It is done by addition of hydrazine, sodium sulphite or sodium sulphide to boiled water.

a)Hydrazine reacts with dissolved oxygen of boiler water forming nitrogen and water.

$3N_2H_2 \longrightarrow 4NH_3 + H_2$

On the other hand both sodium suphite or sodium sulphide reacts with dissolved oxygen to form sodium sulphate which under high pressure decomposes to SO_2

$2Na_2SO_3 + O_2 \longrightarrow 2Na_2SO_4$

$Na_2S + 2O_2 \longrightarrow Na_2SO_4$

The SO_2 penetrates into steam pipes to form corrosive sulphurous acid (H_2SO_3).So a very low concentration of 5 -10ppm of Na_2SO_3 in the boiler water is maintained.

Mechanical deaeration - For removal of dissolved oxygen mechanical deaeration is followed.The process comprises of spraying water over perforated plate fitted tower, heated from sides and connected to vacuum pump.Following Henry's law-high temperature, low pressure and large exposed surface (perforated plates) reduces the solubility of the gas and hence helps in removal of dissolved oxygen from water.

Short Answers Questions-

1. What are major sources of water?
2. Why rain water is the purest form of water?
3. What is hardness of water?
4. What is the reason for hardness of water?
5. Distinguish between temporary and permanent hardness of water?
6. Why does hardwater consume a lot of soap?
7. What type of ligand is EDTA?
8. Why the complex of Ca^{+2} or Mg^{+2} with EDTA is more stable than with EBT?
9. Define the cause of alkalinity?
10. What is the main difference between cold and hot lime soda process?
11. How silica can be removed from water?
12. Why coagulants are not used in hot lime soda process?
13. What is zeolite?
14. What is the difference between zeolite and permutit?
15. Write the principle of zeolite softener?
16. Is the water obtained from zeolite process is free of all impurities including sodium?
17. Give the principle of ion exchange softness?
18. How zeolite is regenerated?
19. How ion-exchange resins are regenerated?
20. That is the difference between sedimentation and coagulation?
21. What do you mean by disinfection in water?
22. Boiled water is not 100% safe drinking water .Explain?
23. What is Reverse osmosis?

Long Answer Questions -

1. What is hardness of water? Why some water are hard?
2. How hard-water is softened by lime-soda process?Explain with chemical reaction.
3. What are the disadvantages of hard-water?How it is removed by ion-exchange method?
4. How hardness of water is estimated by EDTA method?
5. What do you mean by external and internal treatment of hardwater softening?
6. Write short notes on -

   i)Phosphate Conditioning

   ii)Calgon Conditioning

   iii)Carbonate Conditioning.
7. What is caustic embrittlement? What are the factors responsible for it?
8. Write a comparative study of hardwater softening by lime soda process and ion-exchange process?

9.Distinguish between-

i)Purification and softening of water

ii)Cold lime soda process and hot lime soda process.

iii) Softening and demineralisation of water.

10. What do you mean by disinfection of drinking water supply? What are the chemicals used for the process.
11. Why hot lime soda process is better than cold lime soda process?
12. What do you mean by reverse osmosis? How it helps in desalination of water,What are its advantages?
13. Describe the principle of zeolite softening process for hardwater along with its merits and demerits?
14. What is sterilisation of water?Explain how it is carried out by using chloride and ozone?
15. Explain the cause and prevention of scale and sludge formation in boilers?
16. Explain why calgon conditioning is better than phosphate conditioning?

Science of Corrosion

Chapter at a glance

- Definition,Types , Theories and Mechanism of Corrosion.
- Chemical or dry corrosion
- Oxidative corrosion
- Corrosion by other gases
- Liquid metal corrosion

Electrochemical or wet corrosion, Galvanic corrosion,Concentration cell corrosion, Drop corrosion,Waterline-corrosion,Pitting-corrosion,Intergranular corrosion,Stress corrosion, cracking

- .Passivation.galvanic Series
- Factors influencing corrosion
- Control and prevention of corrosion, Proper designing ,Material selection, Use of corrosion
- inhibitor,Environment-modification,Cathodic Protection,Anodicprotection,Protective coating

CORROSION AND ITS CONTROL

Corrosion-

Corrosion can be defined as the process of gradual deterioration or degradation of a metal due to chemical or electrochemical reaction with its environment(gaseous or in liquid form).

Generally corrosion is a slow but continuous process that occurs only at the surface of the metal or material(polymer, concrete, etc.) substances that causes deterioration in its properties like:

- Decaying of surfaces of metal/materials
- Weakening of the material due to loss in the cross sectional area.
- Loss of properties like ductility, malleability ,etc.

Cause of Corrosion-

All metals occur in nature at its mineral/ores in the form of its compounds(as oxides, sulphides, sulphates, chlorides, hydroxides , etc.).Metals are extracted from its minerals/ores through reduction process which requires considerable amount of energy. Thus isolated metals are at higher energy state(excited state)than its corresponding ores and minerals .Therefore isolated metals are in thermodynamically unstable state and has a strong tendency to revert to its thermodynamically stable state by forming compounds with its environment (H_2O,O_2, sunlight ,etc).So corrosion can also be defined as the reverse of extraction of metals.

Corrosion (oxidation)

$$\text{Metal} \rightleftharpoons \text{Metal Compounds+ Energy}$$

(Extraction of metals

Or Reduction)

Some very common examples of corrosion are seen in rusting of iron ,tarnishing of silver article or formation of green film on copper article when exposed to atmospheric condition.Usually corrosion or interaction with the environment begins from the surface of the metal thereafter may continue throughout the metal body. The rate of corrosion depends directly on the amount of energy required for extraction of metals from its ore. Although corroded metal is thermodynamically more stable than pure metal but corrosion causes deterioration in its properties like malleability ,ductility, electrical conductivity, etc.

The loss incurred due to corrosionis enormous. Degradation of useful properties of metals by corrosion not only decreases the efficacy of machines leading to loss in production but also contaminates the products. Moreover replacement of the corroded equipment demands both time and capital, which again affects productivity.As per an estimation for iron alone ,about 15%of worlds annual production goes towards replacement of corroded(rusted) iron. Corrosion produces adverse effects on health and environment too. For example unpredictable failure of machinery(due to corrosion)may cause.

1. leakage of hazardous toxic gases and liquids
2. leakage of inflammable gas or liquid from corroded pipeline posing fire hazards
3. contamination of potable water by toxic impurities and so on.

The list is never ending, a complete understanding of detailed mechanism of corrosion is necessary to minimise the adverse effects of corrosion, protect the metals, materials and save the environment.

THEORIES OF CORROSION-

Corrosion reaction are of two types:

I)Direct chemical corrosion or dry corrosion

II)Electrochemical corrosion or wet corrosion

i) Chemical or Dry Corrosion-

Chemical or dry corrosion occurs by dry direct chemical action of metals with atmospheric gases like oxygen, halogen, hydrogen sulphide, sulphur dioxide ,nitrogen or anhydrous inorganic liquids in absence of moisture or water.This process is very sensitive to temperature. The product of corrosion formed get deposited at the point of corrosion on the metal surface itself. This corrosion product may be soluble ,liquid or insoluble. If the corrosion product is soluble or in liquid state it gets removed from the metal surface as it is formed thereby making the metal more susceptible to corrosion. In case of insoluble corrosion product a thin film of the same is formed over the surface which thickens with time. This layer protect the metal from further corrosion as the rate of penetration of the corroded gas decreases and exposure of fresh metal decreases with time.

Dry or chemical corrosion can be classified as follows

a) Oxidative corrosion(corrosion by oxygen)

b) Corrosion by other gases(e.g:Cl_2, H_2, SO_2, F_2,CO_2, H_2S, N_2, etc)

c) Liquid metal corrosion

Chemical Corrosion
- Oxidative Corrosion
- Corrosion by other gases
- Liquid Metal Corrosion

a) **Oxidative Corrosion-**

Oxidative corrosion takes place by action of oxygen(atmospheric or pure) directly on metal surface in absence of moisture at room or elevated temperature.

Alkali and alkaline earth metals rapidly oxidises at low temperature while all other metals excluding gold, silver and platinum(Au,Ag,Pt)require elevated temperature for oxidation.

The chemical reaction for oxidation corrosion is expressed as follows:

$$2M \longrightarrow 2M^{n+} + 2ne^- \text{(oxidation of metals)}$$

$$n/2O_2 + 2ne^- \longrightarrow nO^{2-} \text{ (reduction of oxygen)}$$

$$2M + n/2\,O_2 \longrightarrow 2M^{n+} + nO^{2-} \text{ or } \underset{\text{metal oxide}}{MO} \text{ (total reaction)}$$

Mechanism

The mechanism of oxidative corrosion on metal surface can be outlined as:

At room temperature metal surface absorbs oxygen through physiorption .As the temperature rises the physically adsorbed oxygen undergoes chemisorption, where in metal gets oxidised to metal ions. The electron thus released reduces oxygen to oxide ion and forms metal oxides.

The metal oxide scale formed on the metal surface serves as a barrier between the metal and oxygen, and restricts further oxidation or corrosion of metal.

For oxidative corrosion both metal and oxygen could get in direct contact which can be achieved either by diffusion of the metal outward through the scale or oxygen must diffuse through the scale to reach the metal below.Both type of transfer can occur, but as the size of metal cation is smaller than size of oxide anion the outward diffusion of metal cation is much faster than inward diffusion of oxide ion.Hence corrosion begins with rapid outward diffusion of metal ions.

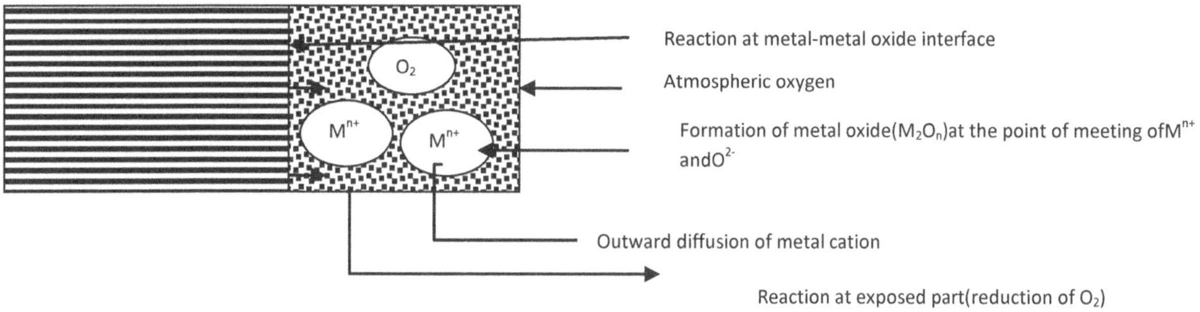

Fig :Oxidative corrosion of metals

This corrosion process is controlled by the nature and thickness of the layer of corrosion product(metal oxide) formed. A layer of metal oxide more than $300A^0$($1A^0=10^{-8}$cm) in thickness is known as scale where less than $300A^0$ is termed as a film. The nature of this film/scale decides continuation of corrosion process as explained below.

i)**<u>Stable film-</u>** A stable film is fine grained in structure and adheres tightly to the parent metal surface. Hence such a layer becomes impervious in nature and prevents inward penetration of oxygen to the underlying metal. Thereby a stable film forms a protective coating over the metal and shields the metal surface against further corrosion.

The oxide films of metals like Al,Cr,Cu,Pb,Sn etc are stable and once formed protects the parent metal from further corrosion.These metals are used in alloy formation with other reactive metal in order to protect the later from corrosion.

ii)**<u>Unstable films-</u>** The oxide layer formed over the metals is unstable as it decomposes back to produce pure metal and oxygen i.e:

Metal Oxide $\rightleftarrows$ Metal +Oxygen

As the process suggests, deterioration due to corrosion does not occur in such metals. Metals like Ag,Au,Pt do not undergo oxidative corrosion.

iii)**<u>Volatile film-</u>**In this case the metal oxide layer formed on the metal surface volatilises out as soon as it is formed making the underlying metal surface susceptible to rapid and continuous oxidative corrosion.The most common example is molybdenum oxide (MoO_3) .MoO_3 being volatile ,Mo metal suffers excessive oxidative corrosion.

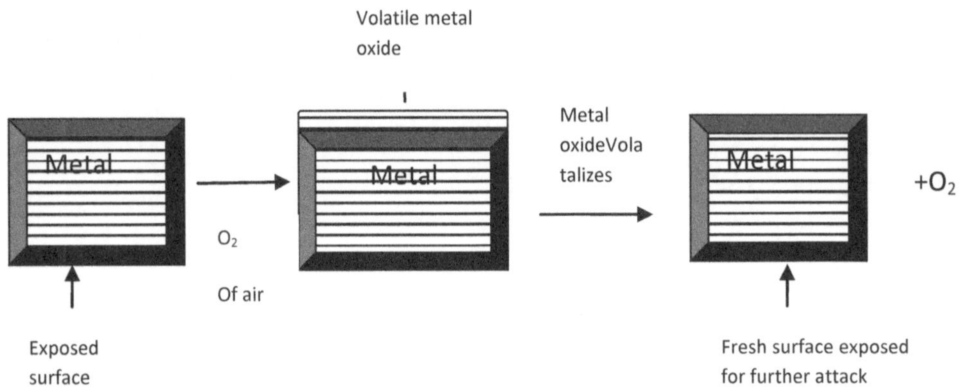

Fig:Volatile oxide layer

$2Mo + 3O_2 \longrightarrow 2Mo_2O_3 \longrightarrow 2Mo + 3O_2$

iv)**Porous film-** Sometime the oxide layer formed over the metal surface is porous and have cracks. As a result oxygen diffuses through these passages and corrodes the metal continuously till the metal is completely converted into its oxides. Alkali and alkaline earth metals suffers this type of corrosion.

Pilling Bedworth Rule:—

Scientist N.B.Pilling and R.E.Bedworth in1923 provided guidelines to determine the nature of the oxide film formed on the metal for the purpose of protection against corrosion. According to Pilling and Bedworth metals can be classified in two Categories: those which forms protective oxides and those that cannot.

The nature of the metal oxide layer formed can be judged from the following ratio known as Pilling-Bedworth Ratio(R_{PB}) as stated below

$$\text{Pilling-Bedworth Ratio}(R_{PB}) = \frac{\text{Volume of metal oxide formed}}{\text{volume of metal consumed}}$$

CASE 1

$$R_{PB} < 1$$

Volume of metal oxide film formed < volume of metal consumed.

In such cases oxide layer formed is consumed in thin and porous(or non continuous).This creates both stress and strain on the oxide layer developing cracks and pores on the surface. Through these cracks and pores oxygen reaches the underlying metal surface and corrodes the metal continuously till the metal is completely converted into its oxides. All alkali(Li, Na,K ,etc) and alkaline earth metals(Ca, Ba, Sr, etc) forms a porous film on its surface.

CASE 2

$$R_{PB} > 1$$

Volume of metal oxide film formed $\geq$ volume of metal consumed.

The thickness of metal oxide layer is equal to or greater than the volume of metal consumed, which makes the layer continuous, adhering and non-porous. This type of oxide film being protective in nature, forms an effective barrier for diffusion of oxygen to underlying layer and retards further corrosion of metal. This type of corrosion layer is found in metals like Al, Cr, Cu,W etc.

b)**Corrosion by Other gases:**

Excluding oxygen metals also get corroded by other gases like SO_2,CO_2,Cl_2,H_2S,F_2 etc.

The extent of corrosion of metal depends firstly on the chemical affinity between the metal and the gas involved and secondly on the formation of protective or non-protective layer on the metal surface.

- If the film formed is protective and non-porous in nature the extent of attack on corrosion decreases.For example Cl_2 gas attacks Ag metal and forms a non porous film of AgCl.

$$2Ag + Cl_2 \longrightarrow 2AgCl$$

On the metal surface.The AgCl film protects the metal from further corrosion .

- If the formed film is non protective and porous in nature, then starting from the surface the whole metal is gradually destroyed.For example Cl_2 gas attacks Sn metal forming volatile $SnCl_4$.

$$Sn + 2Cl_2 \longrightarrow SnCl_4 \text{ (volatile)}$$

$SnCl_4$ filmvolatilises immediately making the fresh metal surface vulnerable to further continuous corrosion.H_2S gas attacks steel forming a porous FeS that makes the underlying metal susceptible to further continuous corrosion.

- H_2S gas attacks steel forming a porous FeS naturally that makes the underlying metal susceptible to further corrosion.

c)**Liquid metal Corrosion:**

This type of corrosion is very common for devices used in nuclear power industry.Liquid metal corrosion occurs when liquid metal is allowed to flow over solid metal at high temperature.This corrosion reaction involves either dissolution of a solid metal by liquid metal or internal penetration of liquid metal into solid metal. Both these modes cause weakening of solid metal by corrosion.

In nuclear power plant sodium metal which is used as a coolant corrodes the cadmium rods.

II) **Electrochemical Corrosion or Wet Corrosion**

Electrochemical or wet corrosion takes place when a metal(or alloy) is in direct contact with either liquid electrolyte or an aqueous solution or even moist air.Electrochemical corrosion occurs with formation of separate anodic and cathodic areas on the same metal(or on two different metal) followed by flow of electric current between the two electrodes using the conductivity solution as medium. Hence the essential condition for formation of an electrochemical corrosion cell are

i)Presence of two electrodes, namely anode and cathode, immersed in an electrically conductive medium.

ii)Existence of electrical potential difference between anode and cathode.

iii)Presence of a metallic path electrically connecting the anode and cathode.

Anodic Area

In anodic area the metal atoms lose electrons to get oxidised to

$$M \longrightarrow M^{n+} + ne$$

metal ions which thereafter either assumes combined state(such as oxides) or dissolves in the electrolyte solution.As a result the anodic metal is destroyed or gets corroded.

M^{n+} ↗ Dissolves in solution

M^{n+} ↘ Forms compounds

Cathodic Area

The electrons released from anodic area passes through the metal and reaches the cathodic area, Although reduction occurs at the cathodic area,the cathode itself does not get affected as most metal cannot be further reduced. It is the dissolved constituents present in the cathodic area accepts the electrons to form ions like OH^- or O^{2-}.

The metallic ions formed in anodic part and non-metallic ions of cathodic area travel towards one another to form a corrosion product in an area between anode and cathode.

Depending upon the nature of the corrosive environment cathodic reaction may proceed either by

i)evolution of hydrogen or ii)by absorption of oxygen.

i) Evolution of Hydrogen

This type of corrosion is usually observed in acidic environment.The electrons released from anode are taken up by the H^+ ions of the acidic solution and eliminates as hydrogen gas.

$$2H^+ + 2e \longrightarrow H_2 \uparrow \quad \text{(in acidic medium)}$$

Rusting of iron in acidic medium can be explained as follows.

Anode - $Fe \longrightarrow Fe^{+2} + 2e$

Cathode - $2H^+ + 2e \longrightarrow H_2 \uparrow$

Therefore the overall reaction is-

$$Fe + 2H^+ \longrightarrow Fe^{+2} + H_2 \uparrow$$

In this type of corrosion the anodic area is larger than cathodes. Similarly all metals present above hydrogen in the electrochemical series have a tendency to displace hydrogen in acidic medium and undergo hydrogen evolution type corrosion.

ii)**Absorption of Oxygen**

As the name suggests the corrosion proceeds through presence of oxygen. Rusting of iron in neutral aqueous solution of electrolyte (like NaCl solution) in presence of atmospheric oxygen a well known example of this type of corrosion.

Surface of iron is usually covered with a layer of iron oxidises which with time develops cracks. Pure iron exposed through these cracks serve as anode while rest of the metal surface behave as cathode. Unlike previous hydrogen evolution type corrosion, here anodic areas are very small compared to large cathodic area.

Pure iron of anodic area gets oxidised with elimination of electrons which reaches the cathodic area through the iron metal, to react with dissolved oxygen present there as represented below:

Anode - $Fe \longrightarrow Fe^{+2} + 2e$ (oxidation)

Cathode- $\frac{1}{2}O_2 + H_2O + 2e \longrightarrow 2OH^-$ (reduction)

Both the ions Fe^{+2} and OH^- diffuse towards each other only to be precipitated as ferrous hydroxide

$$Fe^{+2} + 2\,OH^- \longrightarrow Fe(OH)_2 \downarrow$$

Ferrous hydroxide thus formed gets oxidised to ferric hydroxide in presence of sufficient oxygen.This type of corrosion gets accelerated with increasing supply of oxygen.As an increase in concentration of O_2 shifts the cathodic reaction (reduction) towards right hand side producing more OH^-,which

$$O_2 + 2H_2O + 4e^- \longrightarrow 4OH^-$$

in turn aggravates metal ions to oxidise and supply more electrons.

$$2Fe \longrightarrow 2Fe^{+2} + 4e$$

Thereby increasing the amount of rust formation(i.e; corrosion of metals)

$$2Fe^{+2} + 4OH^- \longrightarrow 2Fe(OH)_2$$

In presence of sufficient oxygen $Fe(OH)_2$ gets oxidised to ferric oxide.

$4Fe(OH)_2 + O_2 + 2H_2O \longrightarrow 4Fe(OH)_3$ or $Fe_2O_3.H_2O$(yellow rust)

With limited supply of oxygen ferrous hydroxide forms black anhydrous magnetite(Fe_3O_4).Rust or hydrated iron(III) oxide has a variable composition of $Fe_2O_3.nH_2O$ .Depending upon the degree of hydration the colour of rust carries from black to yellow or reddish brown.

From above it is evident that although corrosion takes place at anode,"rust" the corrosion product gets deposited at or near cathode. It is because the smaller Fe^{+2} ions diffuse more rapidly towards cathode than larger OH^- ions towards anode.

Galvanic Corrosion-

Galvanic corrosion is an electrochemical process in which when two metals are in conjunction thus electrically connected and exposed to an electrolyte shows preferential corrosion.The metal with lower electrochemical potential acts as an anode and gets corroded while the other metal having higher electrochemical potential acts as a cathode and gets protected.

For example when iron pipes are in conjunction with copper pipes in plumbing shows Galvanic corrosion.The iron portion acts as anode and gets dissolved(corroded)slowly while the copper portion act as cathode and remains unattacked.

Galvanic corrosion is also seen many with bimetallic systems like Zn-Cu system, lead antimony solder around copper wire or with steel screws in a brass marine hardware etc. Hence Galvanic corrosion as also known as bimetallic corrosion.

Mechanism

In acidic solution corrosion follows hydrogen evolution process while in neutral or slightly alkaline solution oxygen absorption takes place.

In Zinc-copper system, zinc being lower in electrochemical potential value(or higher in electrochemical series) than copper it, acts as anode, losses electrons and gets dissolved

$$Zn \xrightarrow{corrosion} Zn^{+2} + 2e$$

Thus electron-current flows from anode to cathode, hence zinc gets corroded and cathodic part copper remains protected.

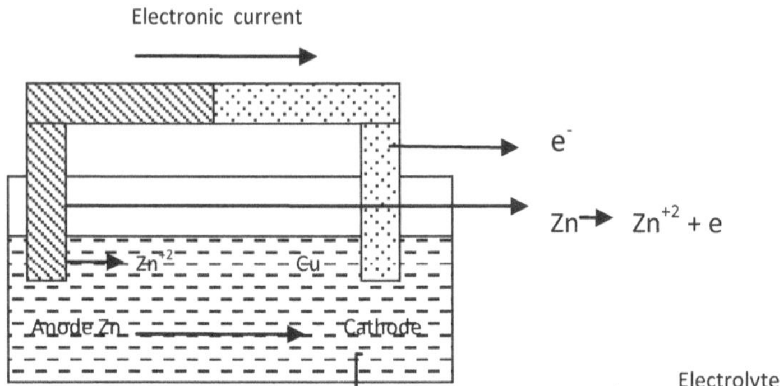

Fig :Galvanic Cell Corrosion

The extent of galvanic corrosion is directly proportional to potential difference between the two metals.

Concentration Cell Corrosion-

As the name signifies this type of corrosion takes place when two or more areas of a metal surface are in contact with different connection of the same electrolyte or due to varying aeration of the same electrolyte. The difference in metal ion concentration happens due to difference in local temperature or slow diffusion of metal ions produced due to corrosion.

Differential Oxygen Concentration Cell:

This is most common type of concentration cell corrosion caused by differential aeration of the metal surface.

Differential aeration occurs when a metal is partially immersed in an electrolyte . In such cases one part of the metal is exposed to air while the other part remains immersed in the electrolyte leading to different air concentration(or aeration) in the two parts of the metal.As a result electrode potential of the metal is different on these two locations.

Under such a situation the areas which are exposed to air ,oxygen becomes cathodic and poorly oxygenated parts become anodic. As a result a differential current flow of anode to cathode leading to corrosion of anodic parts.

Illustration

Let us consider a piece of zinc metal rod partially immersed in a dilute solution of a neutral salt like aqueous NaCl solution and the solution is not stirred.

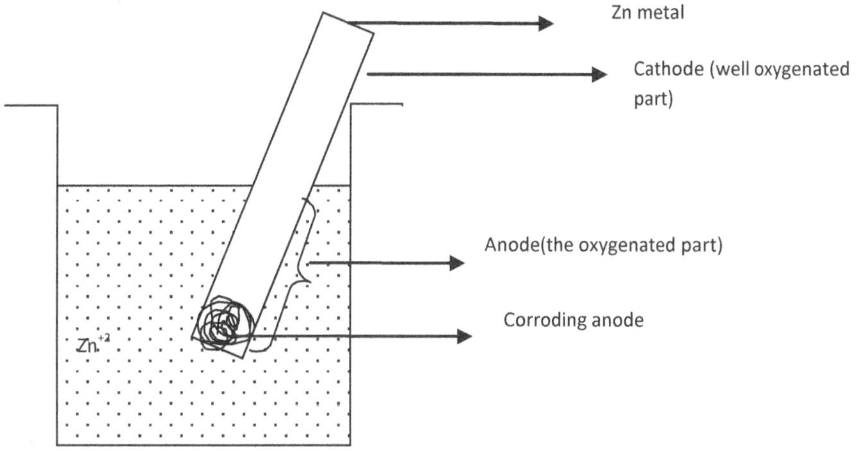

Fig: Corrosion due to differential aeration

The part of the zinc rod above the electrolyte is well aerated (oxygenated) and becomes the cathode. While the other part of the zinc rod under the electrolyte solution has less concentration of oxygen hence poorly oxygenated and acts as the anode.

Potential difference thus develops cause a flow of current between the two areas of the zinc metal as follows

Anode - $Zn \longrightarrow Zn^{2+} + 2e$

Cathode- $\frac{1}{2} O_2 + H_2O + 2e \longrightarrow OH^-$

Thus the migration of ions through the electrolyte and flow of electrons through the metal from anode to cathode completes the cycle.

Drop Corrosion-

This type of corrosion is very common with iron .When drop of water or salt solution falls on iron surface corrosion or rusting is seen in the central region beneath the drop.

This can be explained as the peripheral areas of the drop on the metal have more oxygen compared to the centre of the drop.Thus the metal surface just beneath the centre of the drop becomes anode, while the rest of the metal freely exposed to aeration becomes cathodic.

Anode - $Fe \longrightarrow Fe^{+2} + 2e$

Cathode - $\frac{1}{2} O_2 + H_2O + 2e \longrightarrow OH^-$

The Fe^{+2} moves towards cathode and OH^- ions towards anode and finally combines to form $Fe(OH)_2$ and finally oxidised to rust.

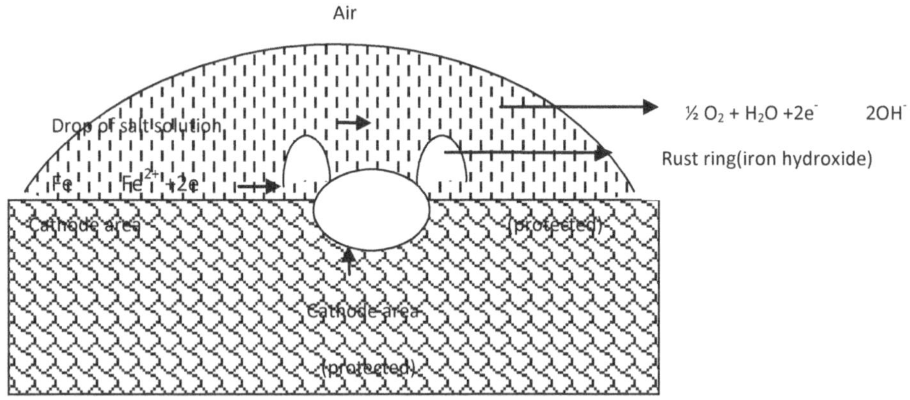

Fig: Mechanism of Drop Corrosion

Waterline Corrosion

This is another form of concentration cell corrosion caused due to differential aeration.This happen when one portion of the base material is submerged in water and other part in air.The reason is same as explained above that the area above the water surface is well aerated while area under the water surface gets less aerated(oxygenated).The differential potential thus develops causing flow of current from anode(less created area) the area under water surface to cathode(well treated area).The area above water surface forming the rust line as explained above .Marine ships are easy victims of waterline corrosion which is controlled by use of special antifouling paints.

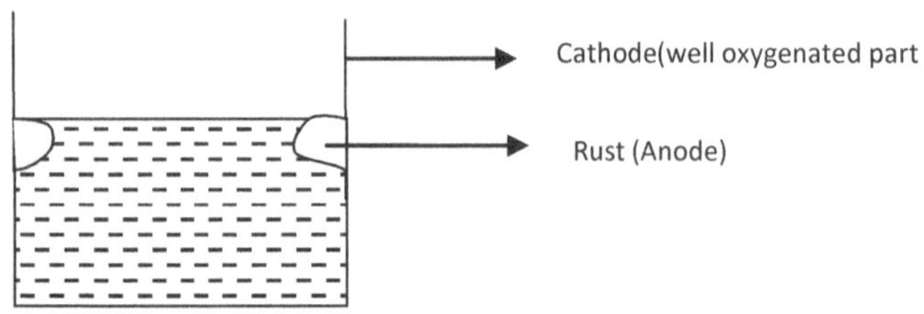

Fig :Waterline Corrosion

Pitting Corrosion-

It is a localised corrosion that results in formation of pits or pin hole like cavities in formation the metal. This type of corrosion occurs due to cracking of the protective film present on metal surface. The driving power for pitting corrosion in the depassivation of a small area ,which becomes anodic while an unknown but potentially vast area becomes cathodic leading to very localised galvanic corrosion.

Thus an anode is formed on the broken film site while the unbroken film(i.e; protective layer) acts as a cathode.The formation of this small anodic and large cathodic areas causes severe corrosion in anodic areas resulting in formation of small pits or holes on the metal surface.

This kind of corrosion is extremely insidious as it causes little loss of material with the small effect on its surface, while it damages the deep structure of the metal. The pits on the surface are often obscured by corrosion product.

Pitting can be initiated by a small surface defect being a scratch or a local change composition, or a damage to the protective coating. Polished surfaces display higher resistance to pitting.

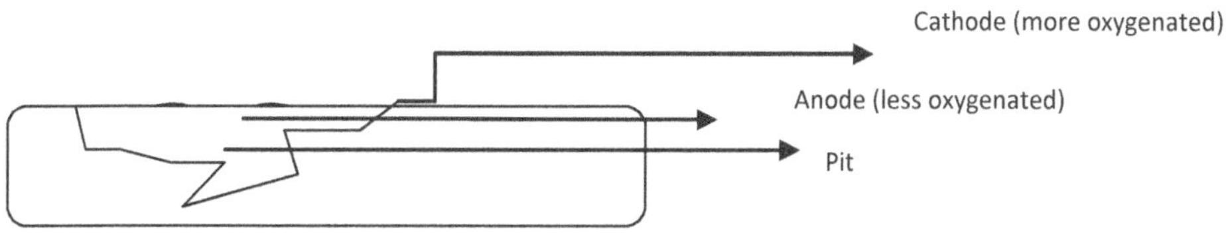

Fig: Pitting corrosion at surface of iron

Reactions:

Anode - $Fe \longrightarrow Fe^{+2}$ +2e

Cathode - $\frac{1}{2} O_2 + H_2O + 2e \longrightarrow 2OH$

Intergranular Corrosion

Intergranular corrosion or inter granular attack is a form of corrosion where the boundaries of the crystallites(grains) of the material are more susceptible to corrosion inner part of the crystallites attack than the phenomenon can be explained as below:

During solidification of molten metal or alloy the process begins with the nuclei which are present in a scattered manner throughout the solid solution. Each of these nuclei grow separately into an atomic array(regular pattern)called grains and the adjacent areas between the grains form grain boundaries .Due to faster precipitation of certain compounds of solid solution at grain boundaries than grain centres a potential difference is created.As a result the grain boundaries become anodic while grain centres behave as cathode. In corrosive environment this potential difference leads to preferential dissolution of anodic grain boundaries leading to dislodgement of the grains due to loss of cohesion.

This type of intergranular corrosion is more prominent in alloys. The most common example is the intergranular corrosion of stainless steel during welding.

Stainless steel is an alloy of Fe, C and Cr. During welding process chromium carbide is precipitated at the grain boundaries. Thus composition of chromium get depleted in region adjacent to grain boundary as compared to grain centre and precipitate.The chromium depleted area becomes anodic and suffers corrosion. The intergranular corrosion occurs on microscopic scale and proceeds along the grain boundaries until the affected grains get completely dislodged due to loss of cohesion.

This corrosion causes sudden failure of machinery without prewarning.

Inter-granular corrosion can be minimised by preventing heterogeneous precipitation through proper heat treatment(that dissolves chromium carbide formed) followed by rapid quenching. Aluminium and copper alloys are also prone to intergranular corrosion.

Stress Corrosion Cracking:

Stress Corrosion cracking is a fracture process which involves a combined and simultaneous action of tensile stress and corrosive environment.

Pure metals are resistant to stress corrosion, only fabricates articles of certain alloys namely zinc brasses, nickel brasses etc, suffers stress corrosion cracking. It is because during fabrication process these materials(alloy) are subjected to uneven stress like rolling, drawing or insufficient annealing. The portion of the material under stress becomes more anodic while rest of the material behave as cathode. Under this situation the presence of a corrosive environment causes the anodic area to suffer corrosion followed by development of cracks. These cracks propagate along a direction perpendicular to applied stress leading to complete breakdown of the material.

Stress Corrosion cracking occurs when the following three requisites coincide

- A susceptible material (metal or alloy)
- Tensile stresses applied to the material
- A specific environment containing an aggressive species that promotes stress corrosion cracking.
- The tensile stress that is applied to the material should be of sufficient magnitude to produce cracking. The corrosive environment is also highly specific. For example an acid chloride solution causes corrosion cracking of stainless steel while traces of ammonia is responsible for stress corrosion cracking of brass.

Example of stress Corrosion Cracking

Season cracking(stress corrosion cracking of Brass)

Brass is an alloy of copper and zinc.In an environment of traces of ammonia or amines, brass suffers stress corrosion.Both the component of brass are soluble in ammonia or amines to form stable complexes i.e; $[Cu(NH_3)_4]^{+2}$ and $[Zn(NH_3)_4]^{+2}$.This dissolution of Cu and Zn in ammonia solution causes fissures which under high tensile stress propagates to form cracks.

Caustic embrittlement of mild steel-

This type of stress cracking is formed in steam boilers and heat transfer instruments.Mild steel when exposed to caustic environment at high temperature and stress, suffers corrosion at crevices near rivets due to caustic embrittlement.Na_2CO_3 is usually added to boiler water as softner, but Na_2CO_3 decomposes at high temperature and pressure producing NaOH.

$Na_2CO_3 + H_2O \longrightarrow 2NaOH + CO_2 \uparrow$

The alkaline water flows through hair cracks and cervices present at inner side of the boiler through capillary action where water evaporates and NaOH concentration build up with passing time the alkali concentration at cracks and crevices increases making alkali concentration in the boiler inside lesser than that present cracks and crevices.This difference of electrolyte concentration generals a concentration cell where the main body containing dil NaOH behaves as cathode and the concentrated NaOH present in cracks and crevices acts as anode.As a result the anodic iron in contact with concentration NaOH get dissolved or gets corroded forming sodium ferrate.The sodium ferrate so formed is further decomposes to magnetic oxide and NaOH causing further corrosion.

$$2Na_2FeO_2 + 4H_2O \longrightarrow Fe_3O_4 + 6NaOH + H_2$$

Prevention-

Caustic embrittlement can be prevented by using

- Sodium phosphate as softening agent instead of sodium carbonate.
- By adding tannin or lignin and Na_2SO_4 as an additive to boiler water blocks the boiler body hair cracks and prevents infiltration of caustic soda.

Passivation or Passivity-

Passivation or passivity is the phenomenon in which a metal or alloy shows much higher resistance to corrosion.It is due to formation of a highly protective, very thin($\sim$ 0.0004mm thick) and quite invisible surface film on metal or alloy.

Passivation of a metal or alloy takes place only in certain environmental condition. For example metals like Ti,Al,Cr and a wide variety of stainless steel alloys containing Cr forms a protective surface film in oxidising environment.The film is also self healing in nature and automatically repairs for any damage that occurs in it in oxidizing environment.But in reducing environment the metal becomes quite active and gets corroded.

Galvanic Series :

In an electrochemical series element are arranged in ascending order of their reduction potential. Hence the metal present higher in the series is more anodic than its lower members, and corrodes faster. However due to passivation a very thin protective film forms on some metal and later becomes more positive and act cathodic and exhibits noble behaviour.As a result the metal behaves differently than as predicted by electrochemical series.Hence a more reliable series based on actual corrosion behaviour of metals and alloys in a given environment is prepared and named as **'Galvanic Series.'**

ACTIVE

Mg, Mg alloys, Zn ,Al, Cd, Duralumin,Mild Steel, Cast Iron, High Ni cast iron,Pd-Sn Solder, → ,Ni, Bronze, Cupro nickel alloys, Cu and Cu alloys,Iconel ,Sn Pb ↓ PASSIVE ← ↓ Stainless steel,Ag,Ti,Graphite,Gold,Platinum →

Noble

Table :Galvanic Series.

Factors influencing Corrosion :

The main factors responsible for corrosion of a metal can be broadly classified as follows:

- Nature of metal
- Nature of the environment

Nature of metal: The rate and extent of corrosion a metal suffers, depends on several factors which are explained below serially.

i)Position in the Galvanic Series

When two metals or alloys are in electrical contact in presence of an electrolyte then the metal or alloy higher up in positioning Galvanic series becomes anodic and suffers corrosion.The greater the difference in position of the two metals ,the higher is the extent of corrosion of the anodic metal.

ii)Purity of the metal

Rate of corrosion is directly proportional to the extent of impurities present in the metal.The impurities present in metal generates heterogeneity cause formation of minute electrochemical cell that accelerates the corrosion of anodic part of the metal.As an example pure zinc is not corroded in very pure hydrochloric acid, but zinc in presence of impurities like Pb or Fe undergoes fast corrosion.The rate of extent of corrosion depends upon exposure and extend of impurities present.A pure metal is usually resistant to corrosion, but pure metals are very expensive and lacks strength hence not suitable to be used in engineering material.

iii)**Physical state of the metal**

The rate and extent of corrosion is also influenced by physical state of the metal. By physical state of metal it includes orientation of crystals, grain size, grain structure, localised stress, scratches etc.For example under corrosive environment grain boundaries are attacked more rapidly than grain centres similarly smaller than grain size of the metal or alloy greater will be the solubility than macroscopic crystals and higher will be the corrosion Or for areas under stress(even for pure metals)becomes anodic and suffers corrosion.

iv) Nature of the oxide film:

Most of the metals undergoes aerial oxidation and get covered by an oxide film.The nature of this oxide film formed is calculated as:

$$\text{specific volume ratio} = \frac{\text{volume of metal oxide formed}}{\text{volume of metal}}$$

If the specific volume ratio >1 ,then the metal is resistant to corrosion.

If the specific volume ratio <1 ,metal is susceptible to corrosion.

Or, other words when the specific volume ratio is less the oxide film formed on the metal is porous which allows oxygen through it and protects the metal from corrosion.For example: alkaline earth metal like Be,Mg,Ca have specific ratio volume low thus these metals easily get corroded while W,Cr,Ni are least susceptible to corrosion as their specific ratio values are 3.6,200 and 1.6 respectively.

v) Area Effect:

The extent of corrosion is also controlled by the relative area of cathode and anode.If the area of cathode is large enough compare to anode then to meet the demand of more electrons for cathode area, the anode undergoes brisk, rapid and severe corrosion.For this reason in copper plates(cathode) having steel revets (anode) exhibits complete corrosion of steel rivets (small area).

vi)Solubility of the products of corrosion :

The solubility of corrosion product plays an important role in deciding the extent of corrosion.If the product of corrosion is soluble in the corroding medium then corrosion proceeds at a faster rate, but if the corrosion product remains insoluble or forms a secondary product with corroding medium which act as a corrosion barrier and protects the metal from getting corroded.

vii)Volatility of Corrosion products :

If the corrosion product is volatile it leaves the metal surface as soon as it is formed and thereby exposing the metal surface to further corrosion .As it is found with oxidative corrosion of Mo by formation of volatile MoO_3.

$$2Mo + 3\,O_2 \longrightarrow 2MoO_3$$

NATURE OF THE ENVIRONMENT

Nature of the environment in and around the metal is an important parameter for its corrosion which can be numbered as below:

i)Temperature- It has been proved from Arrhenius theory of reaction becomes double by an increase of $10^{o}$of temperature, similar rate of corrosion also increases with increase in temperature.Higher temperature not only fastens the rate of corrosion but also passive metals becomes active and gets corroded.

ii)**<u>Humidity-</u>** Usually the rate of corrosion is accelerated with increase of humidity of the environment.The humidity above which the rate of corrosion increases sharply is known as " relative humidity".It is because

- Humidity(moisture) act as a solvent for O_2,CO_2 and H_2S etc. present in the atmosphere and helps in settings the electrochemical cell.
- Dissolution of NaCl in water increases the conductivity thereby increasing the rate of corrosion.
- Metal oxide films present on the metal absorbs moisture and enhance rate of corrosion.
- Rain water also sometime washes away the metal oxide film present on the metal, thereby exposing the metal for further corrosion.

iii)**Effect of P_H-** Corrosion is most effective in acidic medium, followed by alkaline and neutral medium respectively.Thus lower the P_H value greater is the corrosion of the metal. Thereby corrosion can be decreased by increasing the P_H value of the medium.

Zinc the most corrosive metal gets corroded by Carbonic acid the weakest acid ,Its corrosion can be minimised at P_H 11, with further increase of P_H Zn forms a complexion and becomes soluble.

Amphoteric metals like Pb,Al etc dissolves in alkaline solution forming complex ions.

iv)**Nature of ions present-** The nature of the ions present in electrolyte affects corrosion as follows;

- Anions like silicates form insoluble silicates(gel) and prevents further corrosion. On the contrary chloride ions destroy the protective film and expose the metallic surface for further corrosion(e.g: Corrosion of Al in saline water).
- Presence of cyanide ions causes stress corrosion cracking of steel.

Iron undergoes rapid corrosion in presence of NH^+_4 ions.

v)**Conductance of the corroding medium-**For corrosion of underground or submerged area conductance of the medium plays an important role. Dry sandy soils have lower conductance compare to clayey or unrealisedsoil.So strong currents from power leakage cause more damage to metallic structure buried under clayey mineralised soils than those under dry sandy soils.

vi)**Formation of oxygen concentration cell-** Corrosion due to differential aeration (oxygen) has already explained earlier.However the rate of corrosion increases due to differential aeration.The less oxygen concentration part behaves like anode and gets corroded while the most oxygenated part as cathode.This type of corrosion is common in

- metal washers where oxygen cannot diffuse
- underground pipelines passing through different types of soils
- crevice corrosion and waterline corrosion.

vii)**Presence of impurities in atmosphere-**Atmospheric corrosion increases with presence of pollutants like SO_2,H_2S,H_2SO_4,HCl, smoke, dust particle(common in the vicinity of industrial area)

$NaCl$ and$(NH_4)_2SO_4$ present in air near marine atmosphere along with moisture increases conductivity ,resulting corrosion of the metal.Presence of suspended particle in air due to smoke or dust inscreases adsorption of SO_2 and moisture causing corrosion of metals and alloys.

CONTROL AND PREVENTION OF CORROSION

After studying the factors influencing corrosion, it is possible to prevent or even control corrosion and protect the materials from corroding by observing the following guidelines based on above influencing factors:

1. Proper designing
2. Material selection
3. Modification of Environment
4. Modification of properties of metal
5. Use of inhibitor
6. Use of protective coatings
7. Cathodic or electrochemical protection

1.Proper designing

For prevention of corrosion proper designing of the material is the prime important factor.This is summarised as follows;

- If two dissimilar metals are in contact, they should be positioned as close as possible in the galvanic series, otherwise the more active metal will suffer corrosion.
- Direct contact between two dissimilar metal should be avoided.In situation where direct contact is unavoidable ,an insulating fitting should be inserted in between two metals to avoid the direct contact between them.For example plastic washers are used in place of metallic one.
- As the corrosion rate depends on relative size of anode and cathode, the anode should have larger area compared to cathode to control corrosion.
- If two dissimilar metal are in contact then the anode should not be covered with painting or coated as a tiniest break or crack will result in localised corrosion of the material.
- Presence of any crevices between adjacent parts should be avoided as the inhomogeneities develop concentration difference between the metals and causes corrosion.For this reason welded butt joints between same or different metal/alloys are preferred over bolting or riveted
- While designing, proper care must be taken to avoid sharp corners and recesses as dust and moisture accumulates there leading to corrosion due to differential aeration.

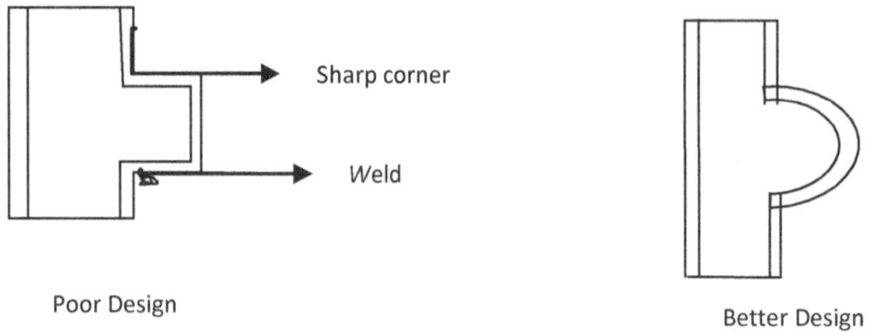

Fig:Proper designing

- To avoid moisture accumulation (leading to corrosion)the equipment should be properly design to facilitate complete water drainage.

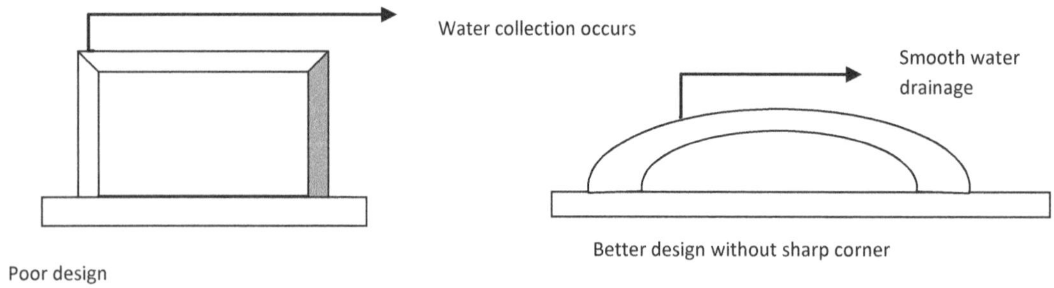

- Equipments should be placed on stands to facilitate free air circulation and avoid water accumulation.
- The anodic parts should be made thick or readily replaceable for increasing life expentancy of the equipment.
- Proper designing of an equipment is necessary as it prevents condition subjecting some areas of structure to stress, that may set up a galvanic couple with non stressed area of the metal/alloys leading to corrosion.

2.**Material Selection**

Another important factor for prevention of corrosion is the type of material(metal or alloy) is used.

- It has already been discussed that impurity of a metal causes heterogeneity followed by corrosion.So heterogeneity of a metal can be decreased by increasing purity of the metal, or in otherwise corrosion resistance of a metal increases with increase in purity of a metal, but pure metal can not be used everywhere due to following reasons

  -Pure metals are usually soft and lacks mechanical strength

  - Pure metals are expensive too

- Corrosion resistance of metals can be increased by alloying with suitable elements.The alloys may be homogeneous or heterogenous.

i)**Homogeneous alloys** are solid solution of two or more metals forming a uniform homogeneous material, eg: Stainless steel which is homogeneous alloy of iron and Cr.Stainless steel containing 13% Cr is used as cutlery, surgical instruments but with 13 to 25% Cr it is in terbine brackets, heat resistant parts etc.

ii)**Heterogeneous alloys** -These are mixtures of two or more components that exits as separates phases.However homogeneous alloys are better then heterogeneous ones as galvanic coupling is absent in the former case.

3.Use of Corrosion inhibitor

Corrosion inhibitors are chemical substances with added in small quantities to corrosion environment(corrodants) decreases the rate of corrosion.Depending on the use the corrosion inhibitor may be anodic, cathodic or vapour phase.

Anodic inhibitors- Alkalines, molybdates, chromates, phosphates ,tungstates or other ions of transition elements with high oxygen content behave like anodic inhibitors.These ions when mixed with corrodants attacks the ions of anode to insoluble precipitates.These precipitates are adsorbs on the metal surface, forming a barrier as protective film and reduce the corrosion rate. However the method has a severe drawback that if due to any leakage or insufficient concentration of protective layer any tiniest part of the anodic metal is left uncovered ,it suffers severe localised corrosion.

Cathodic inhibitors- The main cathodic reaction in acidic medium can be represented as follows:

Equation: $2H^{+}_{(aq)} + 2e \longrightarrow H_2(g)$

Thus by slowing down the diffusion of hydrated H^+ ions to the cathode and/or by increasing the over voltage of (H_2) evolution corrosion rate can be reducd.The diffusion of hydrogen ions (H+)is considerably reduced by addition of inhibitors such as amines, mercaptans, heavy metal soaps, substituted ureas and thiourea.These inhibitors adsorb over the cathodic metal surface and act as a protective layer.Antimony and arsenic oxides deposits adherent film of metallic arsenic and antimony at the cathode area, thereby increasing the over-voltage of hydrogen.

Neutral Inhibitors- The cathodic reaction in neutral solution is represented as:

$H_2O + 1/2\ O_2 + 2\ e^- = 2\ OH^-$

The formation of OH^- is due to the presence of oxygen, thus corrosion can be controlled either by removing oxygen by de aeration or by additive of reducing agents like Na_2SO_3 or by retarding the diffusion of O_2 to cathodic areas. The later is achieved by addition of Zn,Ni,Cr or Mg salts.These salts react with OH^- at cathode to form insoluble hydroxides which get deposited at cathode forming more or less impermeable self barriers.

Vapour Phase Inhibitor- Organic inhibitors like dicyclohexylammonium nitrite, cyclohexyl amine carbonate which readily vaporise and form a protective layer of inhibitors on the metal surfaces are used to prevent corrosion in enclosed spaces or during storage.The vapour phase inhibitor(VPIs) are usually used for protection of sophisticated machinery/equipments.

4.Modification of Environment

The corrosion rate can be controlled by modification of the environment.This can be achieved either by i) removal of harmful corrodents or ii) By deactivating it by neutralisation.

a)**Deaeration** :- In oxygen concentration type corrosive oxygen is removed from the system by temperature adjustment along with mechanical agitation, Along with O_2, the CO_2 content of water also get reduce. Thus the corrosion rate decreases.

b)**Deactivation**:- In this method chemicals like hydrazine sodium sulphite are added to the corrosion mixture

$$N_2H_4+O_2 \longrightarrow N_2+2H_2O$$

$$2Na_2SO_3 +O_2 \longrightarrow 2Na_2SO_4$$

These chemicals rapidly combine with oxygen in aqueous solution as explained.

c)**Dehumidification-** Reduces the moisture content of air by adsorption and resits corrosion of metals. Silica gel, alumina are common dehumidifier used for the purpose.

d) **Alkaline Neutralisation** :- This method is very popular for controlling corrosion of refinery equipments. In this method corrosion due to acidic character of the corrodent is neutralised with alkaline neutraliser(like NH_3, NaOH ,lime etc) injected into liquid or vapour phase.

5. Cathodic Protection or Electrochemical Protection

The principle of cathodic protection is to force the metal to be protected to behave like cathode, as a result the metal does not get corroded. Cathodic protection can be achieved in following two ways:

i)Sacrificial anodic protection(Galvanic protection)

ii)Impressed current cathodic protection

i)Sacrificial anodic protection(Galvanic protection)

In this method the metallic structure(to be protected) is made cathode by connecting by a wire to a more anodic metal.As a result corrosion

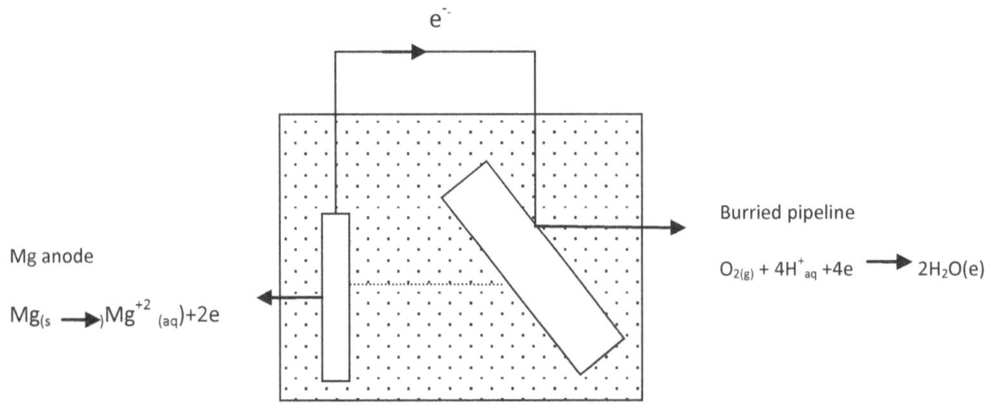

Fig:Galvanic protection using sacrificial anode

takes place at the more anodic metal and parent structure(cathodic) gets protected. Metals used as more active anode are usually Mg, Zn, Al and their alloys.The reaction takes place as follows:

At anode-equation $Mg_{(s)} \longrightarrow Mg_{(aq)} + 2e$

At cathode- equation $O_2 + 4H^+_{(aq)} + 4e \longrightarrow 2H_2O$

The more active metal used is known as sacrificial anode.The sacrificial anode is to be replaced time to time when it gets completely corroded or consumed.

Cathodic protection method is used mainly for protection of underground cables, pipelines and for marine structures (ship hulls)etc. In domestic water boilers, magnesium rods are inserted as sacrificial anode.

ii)Impressed current cathodic protection

The principle of the method is to nullify the corrosion current by application of an impressed current in opposite direction so as to convert the corroding metal from anode to cathode.

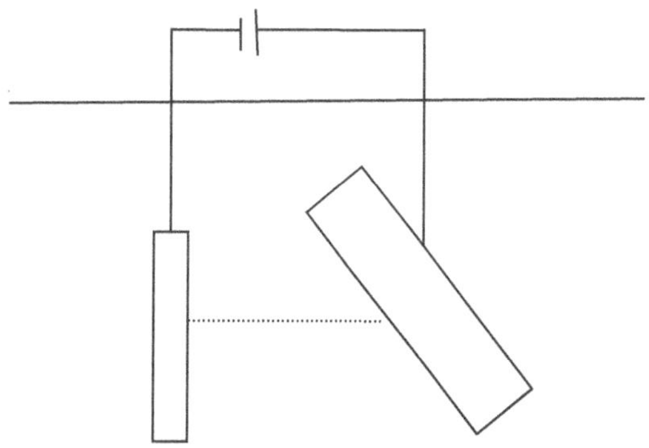

Fig: Impressed current cathodic protection

The corroded metal is made cathode and connected to the negative terminal of DC source while the positive terminal is connected to an insoluble anode like graphite, scrap iron or platinum.Usually sufficient DC current is applied to the insoluble anode (buried in soil/corroding medium) to cathode the metallic structure just in opposite direction of corrosive current.This type of cathodic protection is applied to marine piers, buried pipelines(conducting water or oil) ,condensers ,transmission line towers.

Anodic Protection:-

It is a technique of controlling the corrosion of a metal surface by making it anode, of an electrochemical cell and controlling the electrode potential in a zone where the metal is passive.Under these condition the metal surface is passivated by deposition of oxide film, and the rate of corrosion reduces.This method is used for steel, stainless steel, aluminium and chromium.

6. Protective Coatings

Protecting the surface of the material with protective coating in one of the primitive methods to resists corrosion.A coated surface isolates the underlying surface from the corrosion provided the applied coating must be chemically inert with the environment at particular temperatures and pressure.Secondly the coating must prevent penetration of environmental material which they protect.Coating used to control corrosion may be Metallic or Non-metallic.

Metallic Coating- Metals like Ni,Zn,Sn,Cd and Cr capable of forming oxide films that adhere to base metal surface and prevent corrosion are used for coatings. Coatings can be applied by various techniques.

i)**Electroplating-** In this method the metal to be plated(base metal) is made the cathode while anode is made of coating material or inert material.The electrolyte contains the ions of the coating material, with passage of electric current the base metal gets electrocoated with the coating material.

ii)**Metal Cladding-** In this method the base metal is sandwiched .between two layers of protecting metal and compressed through rollers under action of heat and pressure.The cladding materials are itself corrosion resistant like Ni,Cu,Ag,Pb and alloys also.

iii)**Dipping-**The base metal is dipped in bath of molten coating metal.The thickness of the coating is adjusted by squeezing through rollers galvanization of steel(i.e; coating of zinc on iron) is made by this method.

iv)**Spraying-** It is one of the modern techniques where a stream of atomised droplets are blown on base metal surface.

Non-Metallic Coating

This maybe inorganic or organic coating on base metals.

Organic Coating- Organic coating on metal is an inert organic barrier to protect the metal against corrosion.Organic coating include- paints ,varnishes, enamels, lacquers etc.

Chemical Conversion Coating- These are inorganic barrier against corrosion attack on base metal. This is applied through chemical or electrochemical process.These coatings provide protection and add aesthetic value to base metal.Chemical coatings like phosphate coating, chromate coatings etc. belong to these category.

QUESTIONS

Short questions

1. What is corrosion?
2. What is rusting of iron?
3. What is Galvanic corrosion?
4. Where electrochemical corrosion takes place at anode or cathode?
5. what is wet corrosion?
6. What is electrochemical corrosion?
7. Iron corrodes faster than Al although Fe is placed below Al in electrochemical series.Explain?
8. What do you known by the term Drop corrosion?
9. What is waterline corrosion?
10. Write a note on season cracking?
11. What do you know by stress corrosion cracking of brass?
12. What is passivity?
13. Write howP_H of the medium effects corrosion?
14. Why a wire mess corrodes faster at joints?
15. Why in identical condition an impure metal corrodes faster than pure metal?Explain
16. What is cathodic protection?
17. Name the volatile oxidative corrosion product of a metal.
18. What is differential aeration corrosion?
19. Why nut and bolt made of same metal is preferred?
20. Differentiate the corrosion of Al and Mg?
21. Smaller the anodic area greater is the corrosion. Justify the statement
22. What is the effect of grain size on corrosion?
23. Why chromium used as an alloying metal for steel and iron?
24. Explain the effect of temperature on metallic corrosion.
25. What do you understand by the term selective leaching?
26. Name one volatile oxidative corrosion product of a metal(MoO_3)
27. Name some metals whose specific volume of their oxides is greater than the metal(Cr,Ni,Al,Ti)
28. Why rusting of iron is quicker in saline water than ordinary water?

29. When in a structure two different metals are in contact which metal will corrode?

30. What is the requirement of the coating metal to act as sacrificial anode for iron?

Long Questions

1. Why do some metal corrode?
2. Differentiate between chemical and electrochemical corrosion?
3. Discuss about the mechanism of oxidative corrosion?
4. What is Pilling-Bedworth rule?
5. Write a note on concentration cell corrosion?
6. What do you known about Pitting corrosion?
7. What is waterline corrosion?
8. What is drop corrosion?
9. What is inter-granular corrosion?
10. Write a note on galvanic corrosion?
11. What are the factors that influence corrosion?
12. Write a short note on control and prevention of corrosion?
13. How the use of corrosion inhibitor check corrosion of metals?
14. Between Galvanic series and electrochemical series which one is preferred for prediction of corrosion of metals and why?
15. Write a note on cathodic protection?
16. What do you mean by impressed current cathodic protection?
17. How corrosion can be controlled by proper designing of the equipment?.

Fuels

Chapter at a glance

- Introduction & classification
- Calorific Value
- Analysis of coal and Coke
- Refining of Crude Oil
- Fractional distillation
- Cracking
- Knocking and anti-knocking
- Octane and Cetane number
- LPG and CNG
- Combustion Calculation
- Synthetic petrol
- Refining of gasoline
- Use of diesel, kerosene and LPG Gas fuel
- Non-petroleum fuels. (Natural gas, water gas, oil gas, producer gas, bio-gas).
- Analysis of flue gas
- Questionss

Introduction

Fuel is recognised as the substance which provides energy. Energy can be generated by burning fuels, from nuclear reactions or from other alternative sources.

Thus fuels can be defined as substances which on combustion or reaction with other substances produces large amount of energy (mainly heat energy) that can be economically used for industrial and domestic purposes.

Combustion of fuel can be represented as:

$$\text{Fuel} + O_2 \rightarrow \overset{\text{Product}}{(CO_2, HO_2\text{etc})} + \Delta H$$

$$\Delta H = E_R - E_P$$

Where ΔH = Heat Energy

E_R =Energy of reactants

E_P =Energy of products

The amount of heat generated by a fuel depends mostly on the carbon and hydrogen content of the fuel. Primary source of fuel is fossil fuels like coal, mineral oil and natural gas. The non-carbonaceous substances like boron hydride produces enormous amount of heat energy on burning with air or oxygen. It is mixed with hydrogen peroxide and being used as a rocket fuel.

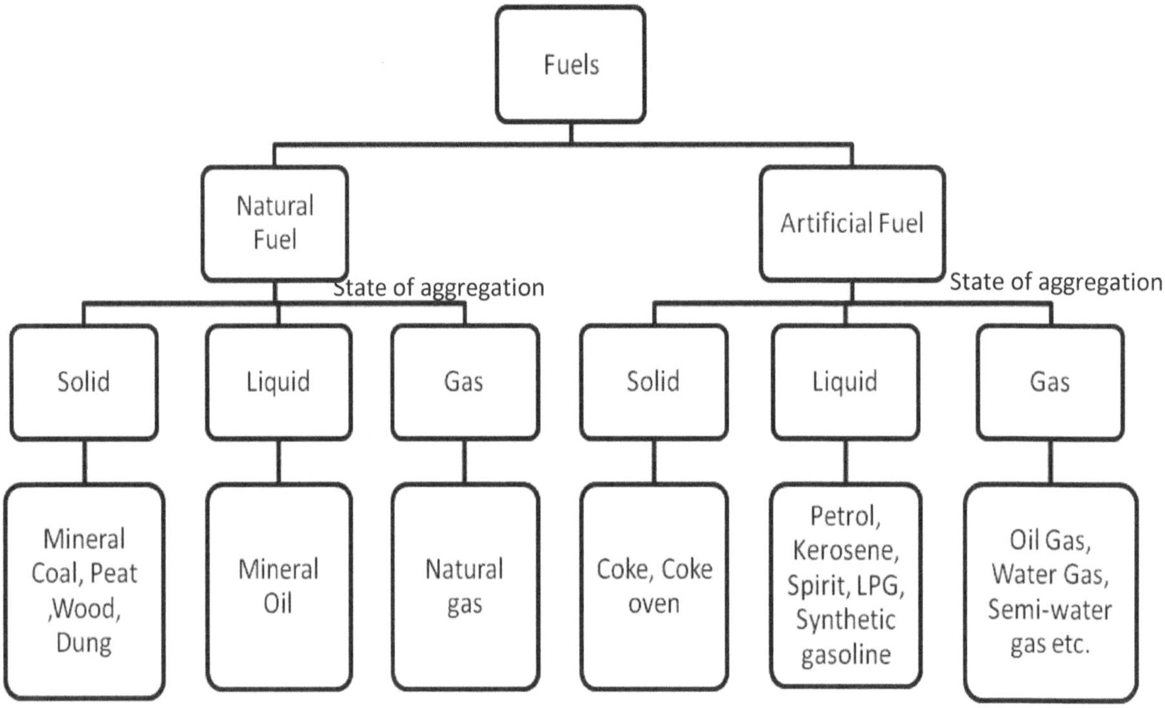

Classification

Depending on the i) occurrence and ii) state of aggregation the carbonaceous (fossil) fuels are classified as follows:

Colloidal fuel

It is an artificial fuel consists of a suspension of finely powdered coal in liquid fuel (e.g. liquid oil) in the ratio 3:2 by weight. The colloidal suspension is maintained by the addition of a stabilizing agent to it. The advantages of colloidal fuel are

- Colloidal fuels possess high calorific value.
- Comparatively easy to handle than powdered coal.

- Colloidal fuel used as an industrial fuel for cement, brick, ceramic and metal processing industries.

Calorific Value

Calorific value of a fuel is defined as the total quantity of heat liberated by complete combustion of unit mass (unit volume) of fuel expressed as cal g^{-1} (or kcal kg^{-1}) or is B tu lb^{-1} calorific value of a fuel increases with increase in carbon and hydrogen content of it. Calorific value is a measure of the quantity of a fuel.

Units of heat

1. Calorie – It is the amount of heat required to raise the temperature of one gram of water through one degree centigrade.
2. Kilo Caloric – It is quantity of heat required to increase the temperature of one kilogram of water through one degree centigrade and

   1 KCal = 1000 Cal.
3. British Thermal Unit (B.Th.U) – It is defined as the quantity of heat required to raise the temperature of one pound of water through one degree Farenheit (60-61 ^{0}F). This is English Unit and

   1 B.Th.U = 252 Cal = 0.252 KCal

   1 KCal = 3.968 B.Th.U
4. Centigrade heat Unit (C.H.U) – It is defined as the quantity of heat required to raise temperature of one pound of water through one degree centigrade.

   1 KCal = 2.2 CHU = 3.968 B.Th.U

The efficiency of a fuel is evaluated from its total calorific values which in turn depends on higher or gross calorific valuc (HCV) and lowcr or net calorific value (LCV).

Higher or Gross Calorific Value (HCV)

It is the total amount of heat produced when unit mass (volume) of fuel is completely burnt and the products of combustion are cooled to room temperature (i.e. 15 ^{0}C or 60 ^{0}F).

For instance H, S or N present in the fuel on combustion are converted to water H_2SO_4 ,HNO_3 respectively. On cooling these products of combustion to room temperature, the latent heat of condensation gets included in the measured heat. Hence the name is higher or gross calorific value.

Lower or Net Calorific Value (LCV)

It is lower than HCV. Lower or Net calorific value is defined as the net heat produced when unit mass or volume of the fuel is burnt completely and the products are permitted to escape. Thus

LCV = HCV – Latent heat of products formed (e.g. water vapour).

In actual combustion practice the product of combustion are not allowed to cool but are either allowed to escape into the atmosphere or led into a heat exchanging unit. For instance, in case of hydrogen containing fuels the combustion products like steam is allowed to escape along with hot combustion gases and are not condensed. Hence a lesser amount of heat energy is available than the HCV (theoretical value)

Units

For solid or liquid fuel the unit of calorific value is

Calorie/gram (Cal/g)

or Kilo calorie/kg (Kcal/Kg).

and British Thermal unit/lb (B.Th.U/lb)

For Gaseous fuels the unit is

Kilo calorie/Cubic meter (Kcal/m^3)

B.Th.U/Cubic feet (B.Th.U/ft^3)

Characteristics of a Good Fuel

A good fuel must satisfy the following characteristics:

1. **High calorific value** – A good fuel must posses a high calorific value, since the amount of heat liberated and temperature attained these by depends upon calorific value of a fuel.

2. **Moderate Ignition temperature** – Ignition temperature can be defined as the lowest temperature to which the fuel must be pre-heated so that it starts burning smoothly. Fuels with low ignition temperature are dangerous for storage andtransport fire hazards. But fuels with high ignition temperature although safe for storage, handling and transport but causes difficulty in kindling (ignition). Hence an ideal fuel should have moderate ignition temperature.

3. **High calorific Intensity** – Calorific intensity is the maximum temperature attained when a fuel is burnt. An ideal fuel should have high calorific intensity.

4. **Low moisture content** – For good fuel to have high heating value its moisture content should be low.

5. **Low non-combustible matter content** – An ideal fuel should have very low or no non-combustible matters like ash, clinker after combustion. Presence of non-combustible matters not only reduces the heating value of the fuel but also creates additional cost for storage and disposal of these matters.

6. **Moderate combustion rate** – A good fuel must have a moderate rate of combustion so that it can attain required high temperature. A low rate of combustion may not produce the required high temperature, as a portion of the heat produced is lost due to radiation. On the other hand, a fuel with high rate of combustion becomes difficult to control.

7. **Products of combustion** – A good fuel should not produce harmful substances (CO_x, NO_x, SO_2, H_2S or poisonous gas) nor any gases with offensive odour on combustion.

8. **Cost and Transportation** – A good fuel is one which is easily available and easy to handle for storage and transportation. More over the fuel should be low in cost. For a solid fuel the size should be uniform so that combustion is regular while for liquid and gas fuel combustion should not be spontaneous (moderate ignition temperature).

COMPARATIVE STUDY OF SOLID,LIQUID& GASEOUS FUELS

1. Gaseous Fuels

The advantages of gaseous fuel over liquid or solid fuels are as follows

Advantages

i) Gaseous fuels burn easily

ii) Leaves no residue and produces no or little smoke.

iii) They have high calorific value.

iv) Gaseous fuel gives the highest pyrometric effect.

v) The combustion of gaseous fuels can be controlled easily.

vi) They do not require special burners.

vii) Gaseous fuel can be used for internal combustion engine.

viii) With gaseous fuels the waste heat of combustion may be used for preheating the fuel and air mixture which promote the combustion process.

ix) Gaseous fuels are more economical.

x) Convenience and cleanness are other favourable points.

Disadvantages

i) Gaseous fuels are highly inflammable.

ii) Proper storage facility is essential.

Liquid fuels

Advantages

i) Liquid fuels also burn easily.

ii) Leaves no non-combustible residue (e.g. ash, clinckers etc.)

iii) They have high calorific value.

iv) They have low moisture content.

v) Liquid fuels require less amount of air for complete combustion.

vi) Combustion is easily controllable.

vii) Liquid fuels require less furnace space for combustion.

viii) Liquid fuels can be used for internal combustion engines.

ix) It has moderate ignition temperature.

x) Liquid fuels are economical.

xi) Convenience and cleanliness are two other favourable features.

Disadvantages

i) Liquid fuels are inflammable.

ii) The cost of liquid fuel and its storage is high.

iii) Liquid fuels are not completely combustible.

iv) It may produce bad odour and harmful gases.

Solid fuels

Advantages

i) They have high calorific value.

ii) Solid fuels usually have moderate ignition temperature.

iii) Easy to transport and storage.

iv) Solid fuels are economical.

v) Solid fuels are good adsorbent for gases.

Disadvantages

i) Thermal efficiency is low.

ii) High moisture content.

iii) Require excess air for complete combustion.

iv) Combustion is not easily controllable.

v) Large amount of heat goes waste.

vi) Non-combustible residue like ash, clinckers and harmful gases are produced.

vii) Solid fuels can not be used in internal combustion engines.

DETERMINATION OF CALORIFIC VALUE

Calorific value of fuels can be determined using calorimeters. Different calorimeters are used for solid, liquid and gaseous fuels.

BOMB CALORIMETER

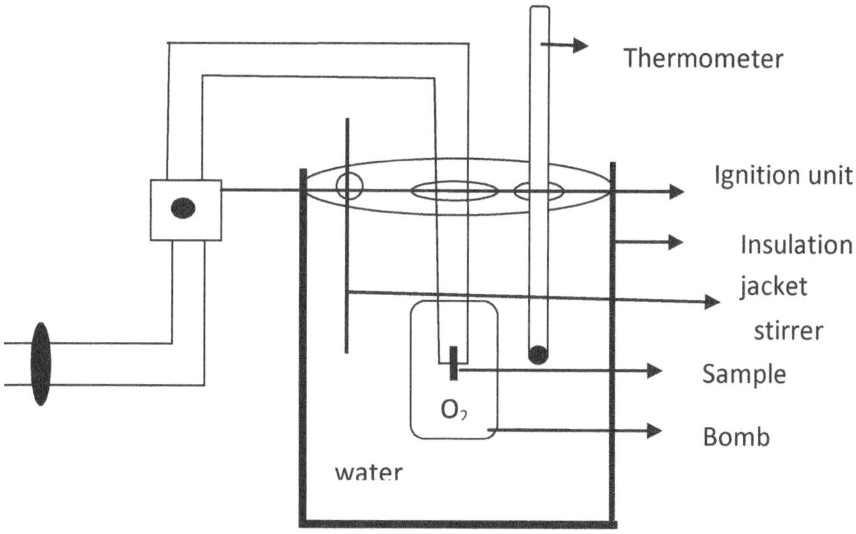

Fig : Bomb calorimeter

Bomb calorimeter is used for determination of calorific values of solid and liquid fuels (hydrocarbons). This apparatus was designed by Berthelot in 1881.

Principle :

$$C_xH_yO_{z(s)} + \frac{\left(2X + \frac{Y}{2} - Z\right)}{2O_2(g)} \rightarrow XCO_{2(g)} + YH_2O_{(l)}$$

As the combustion reactions are exothermic, the heat produced by combustion of a known mass of fuel is absorbed in water and measured. Using the principle of calorimetry, the calorific value of the fuel is calculated

Heat liberated from combustion of fuel = Heat absorbed by water + Heat

absorbed by the apparatus.

A bomb calorimeter consists of a cylindrical stainless steel bomb internally enameled with platinum or some non-oxidisable material. The bomb has a lid made of metal that can be fitted as a gas-tight seal screwing down on a lead washer. The lid is provided with two electrodes and an oxygen inlet valve. To one of the electrodes a small ring is attached to hold a small nickel or stainless steel crucible. The bomb is placed in the copper calorimeter which is surrounded by a double jacket of air and water respectively . The calorimeter is equipped with a stirrer and a Beckmann's thermometer that can detect up to $1/100^{th}$ of a degree.

Function

A known mass (about 0.5 to 1.0 g) of the sample is taken in the crucible and placed inside the bomb. A fine magnesium wire is dipped into the sample and then stretched across the electrodes. The lid of the bomb is screwed tightly and the bomb is filled with oxygen at 25 to 30 atm pressure. The bomb is then placed inside a copper calorimeter with known quantity of water. The initial temperature of water is noted. The electrodes are connected to a 6 volt battery and circuit is completed. The reaction once started progresses rapidly with liberation of large amount of heat. Uniform stirring of water is continued and the final temperature attained is recorded.

Calculation

Let m = Mass of the sample present in the crucible

w = mass of calorimeter system including water

W = Mass of water in the calorimeter

t_1= Initial temperature.

t_2 = Final temperature of water in calorimeter.

L = Higher calorific value in fuel in cal/g.

Total heat liberated from combustion of sample = mL

The heat absorbed by water and system is = $(W + w)(t_2 - t_1)$

: .Heat liberated by the fuel = Heat absorbed by water and apparatus

$mL = (W + w)\ (t_2 - t_1)$

$$\text{or}\ \ L = \frac{(W+w)(t_2-t_1)}{m} = \text{Higher calorific value of fuel} \qquad -(1)$$

LCV = HCV – Latent heat of water produced

= L – Latent heat of water produced

Latent heat of water = 0.09 $\times$ 587 Cal/gm

Latent heat of steam = 587 Cal/gm

Corrections

During the progress of the experiment other processes involving heat change may also occur. So for precision of the result those factors are incorporated.

i) Fuse wire combustion – During ignition of the fuse wire heat is liberated. This amount must be deducted from the heat liberated by fuel value to get the accurate result.

ii) Formation of Acid – S and N present in fuels under high pressure and temperature (of ignition) undergoes oxidation to produce sulphuric and nitric acids respectively as follows.

$$S + 2H + 2O_2 \rightarrow H_2SO_4 + \text{Heat}$$

$$2N + 2H + 3O_2 \rightarrow 2HNO_3 + \text{Heat}$$

These are exothermic reactions. The measured heat value also includes the heat given out during acid formation. Correction is made by deducting the amount of heat evolved due to this exothermic reactions from the measured value. The quantity of the acids formed are analyzed from washings of the bomb.

Amount of H_2SO_4 formed is determined by precipitation with $BaSO_4$. It is found that the correction for

1 mg of S is 2.25 Cal and for 1 ml of $\frac{N}{10}$ HNO_3 formed is 1.43 Cal

iii) Cooling correction

A bomb calorimeter is only approximately adiabatic. In reality a small heat leak may occur through the calorimeter and moreover the stirrer does not work on calorimeter. This non adiabatic condition is corrected with a radiative correction or cooling correction.

The maximum temperature attained by the calorimetric water and the time taken to attain room temperature from maximum temperature by the calorimetric water is noted. From which the rate of cooling (dt^0/minute) and the actual time taken for cooling (t minutes) is computed and the cooling correction dt $\times$ t is added to the rise in temperature.

Thus equation (1) becomes

$$L = \frac{(W + w)(t_2 - t_1 + \text{cooling correction}) - (\text{Acid} + \text{fuse correction})}{\text{Mass of the fuel (s)}} \quad - \quad \textbf{(2)}$$

BOY'S GAS CALORIMETER

Boy's gas calorimeter is designed by Sir Charles Vernon Boys and is used for determination of calorific value of gaseous fuels and easily vaporizable liquid fuels.

Principles:

The gaseous or easily vaporizable liquid fuel is burned at a constant rate and the heat produced is absorbed by water which is circulated in a chimney around the burner. From the volume of the fuel burnt, water collected and increase in temperature of water, the calorific value of the fuel is determined.

Construction:

Boy's gas calorimeter consists of a gas burner in which a known volume of gas can be burnt at a known pressure and at uniform rate of 3 to 4 liters per minute. A combustion chamber or chimney equipped with copper tubing coiled inside as well as outside it, is present around the burner. Water is passed through the coil continuously and at a constant rate. The temperatures of incoming (T_1) and outgoing (T_2) water is recorded with thermometers. During the passage of water it takes away practically whole of the heat of burner and the steam formed during combustion gets condensed and is collected as water in a trough present at the lower portion of chimney. The whole assembly is enclosed in an insulated chamber.

Function:

Initially the calorimeter is warmed up by simultaneous circulation of water and burning of gaseous fuel at constant rate for about 15 minutes. On warming up of the calorimeter the rates of fuel burning and water circulation are adjusted so that the exit water leaves the apparatus approximately at atmospheric pressure. When the steady conditions are established the following reading are taken simultaneously.

i) The volume of gaseous fuel burnt at given temperature and pressure in a certain time period.

ii) The quantity of water circulated through the coil during the same time period.

iii) The change of temperature.

iv) Mass of water condensed during the period

Calculation

Heat produced by burning of gaseous fuel is transferred (absorbed) to water in the copper tubing and the steam formed is condensed back to water that collected in the trough.

Let V = Volume of gaseous fuel burnt at time t.

W = Mass of water circulated at the same time t.

T_1 and T_2 are the initial and final temperature of water respectively.

M = Mass of water condensed in trough at time t

and L = the HCV of fuel in Cal/gm.

Assuming no heat lost

Heat obtained by combustion of fuel = Heat absorbed by circulating water.

$$VL = W(T_2 - T_1)$$

$$HCV = L = \frac{W}{V}(T_2 - T_1)$$

LCV = HCV – Latent heat of water produced

$$= \left(L - \frac{m}{V}587\right) Kcal / m^3$$

Latent heat of water = $\frac{m}{V}587\ KCal/m^3$

JUNKER'S GAS CALORIMETER

Junker's gas calorimeter is also used for determining calorific values of gaseous and easily vaporizable liquid fuels.

Principle

A known volume of gas is burnt in excess of air at a constant rate in a combustion chamber. The total heat produced in the above process is absorbed by the water flowing at a constant rate in the annular space around the combustion chamber. From the change in temperature of water and heat produced during combustion process the calorific value of the fuel used is calculated.

Construction

Junker's gas calorimeter consists of following parts.

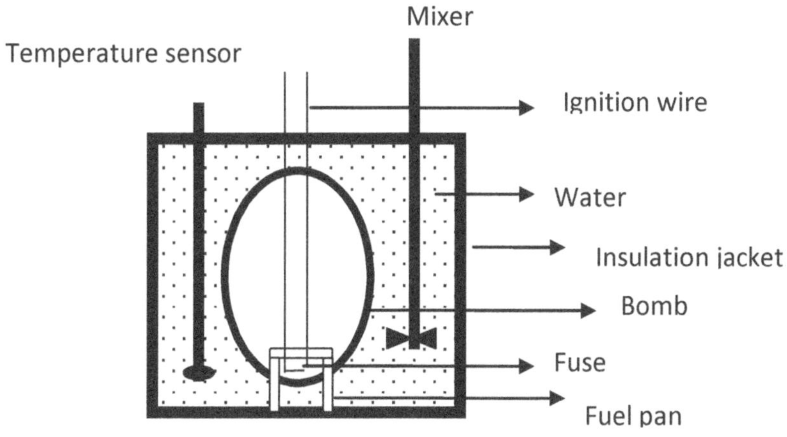

Fig :Junker's gas calorimeter

Bunsen Burner

A special type of bunsen burner is clamped to the calorimeter at the bottom. During the process the burner can be pulled out or pushed into the combustion chamber according to the requirement of process.

Gasometer – This is used to measure the volume of gas burning per unit time. It is fitted with a thermometer and a manometer to note the temperature and pressure of the gas before burning.

Pressure Governor– Pressure governor assists in maintaining the standard condition of saturation of fuel gas with water. Pressure governor regulates the supply of gaseous fuel at a given pressure.

Gas Calorimeter – It consists of a cylindrical combustion chamber surrounded by an annular water space for heating water and fitted with heat exchanger coils. The entire assembly is insulated from loss of heat by radiation and convection with a chromium plated outer jacket. The annular space for water heating is connected at its lower end to internal water supply. Temperature at inlet and outlet of water is recorded with thermometers.

Function – Initially the calorimeter is warmed up by simultaneous burning of gaseous fuel and circulation of water at constant rate for 15-20 minutes. When steady state is attained i.e. the rate of gaseous fuel burning and circulation of water are controlled and combustion products leaves the apparatus at nearly one atmospheric pressure the following reading are noted.

i) Volume of gaseous fuel burnt V at a given temperature and pressure at a certain time period (t).

ii) Quantity of water passing through the annular space in the same time period.

iii) The change of temperature.

iv) Mass of water condensed in the outlet water.

Calculation

Let V = Volume of gaseous fuel burnt at STP in a certain time t.

W = mass of cooling water used in time t.

T_1 = Temperature of inlet water.

T_2 = Temperature of outgoing water.

m = Mass of steam condensed in time t.

Heat produced by combustion of gaseous fuel = VL.

Heat absorbed by water = $W(T_2 - T_1)$.

Assuming no heat loss

$VL = W(T_2 - T_1)$.

$$L = \frac{W}{V}(T_2 - T_1) = H.C.V$$

Mass of water condensed per m^3 of gas $\frac{m}{V}$ kg.

Latent heat of steam per m^3 of gas $= \frac{m}{V} \times 587$ KCal

$$\therefore LCV = \left[HCV - \frac{m \times 587}{V}\right] KCal/m^3.$$

Coal

Coal can be defined as a stratified rock made up of highly carbonaceous matter having fuel value. It contains mainly C, N, H, O, S and non-combustible inorganic matter along with moisture. On careful microscopic examination coal shows presence of aggregation of wood like and cork-like structure confirming its origin from plants.

It is formed in nature as a result of a series of decomposition of vegetable matter under the influence of heat and pressure in a limited supply of air over million of years. The process of coal formation from vegetable debris (plants) can be explained by the underlying two theories as proposed by the geologists.

In situ theory – This theory emphasizes the formation of coal at the site of vegetation itself. Vegetation debris from vegetation grows on wet lands like swamps, bogs and marshes. The high purity of many coal seams is self-explanatory.

Drift (transportation) theory – According to this theory the vegetable debris is believed to have been transported by rivers and streams from its site of growth to places like deep depression, big lakes, huge deltas etc and buried underground with passing of time. These debris under high temperature and excessive pressure underwent anaerobic bacterial transformation to peat with time as follows:

$$2(C_6H_{10}O_5)_n \rightarrow nC_8H_{10}O_5 + 2nCH_4 \uparrow + 2nCO_2 \uparrow + nH_2O$$

The bacterial decay being slow and the cellulosic material underwent decomposition with liberation of methane and carbon dioxide gas. Accumulation of decay products increased the pressure on the debris while the heat of the molten igneous rocks below the seam transformed wood into peat. The peat contains about 60% C, water and volatile matter. The CO_2 released made the environmental acidic and arrests the bacterial action temporarily. However this CO_2 gets absorbed by the alkaline soil and rock and the decay process proceed again. Under combined effect of pressure and temperature the peat deposit buried under sedimentary deposit loses moisture and volatile matter. This decreased the oxygen content while carbon content gets increased and the spongy peat was transformed into hard brittle coal. The process of transformation of peat to coal can be represented as:

Peat $\longrightarrow$ Lignite $\longrightarrow$ Bituminous coal $\rightarrow$ Anthracite coal $\longrightarrow$ Graphite
(Brown coal)

As the world's coal supplies mainly come from regions not having a tropical climate, the drift theory can account for accumulation of organic matter and subsequent transformation into coal.

Ranking/ Classification of coal

The rank of a coal represents its degree of maturity which is primarily based on carbon content of the coal as depicted in table below .

Table : Ranking classification of coal

Type of coal	C	H	O	N	VM	**% Moisture**	**Calorific value Cal/g**
Wood	40-50	5-6	20-40	0-0.5	-	70-90	4000-4500
Peat	45-60	3.5-6.5	20-45	0.75-3	45-75	70-90	4125-5280
Lignite (Brown coal)	60-75	4.5-5.5	17-35	0.75-2	45-60	30-50	6600-7100
Bituminous coal	75-90	4.0-5.5	20-30	0.75-2	11-50	10-20	6600-8800
Anthracite	90-75	3-4	2-3	0.5-2	3.8-10	1.5-3.5	8470-8800

The progressive transformation of wood to anthracite coal results in

- Decrease in moisture content
- Decrease in H, O, N, S and VM (Volatile matter)
- Increase in carbon content
- Increase in calorific value
- Increase in hardness

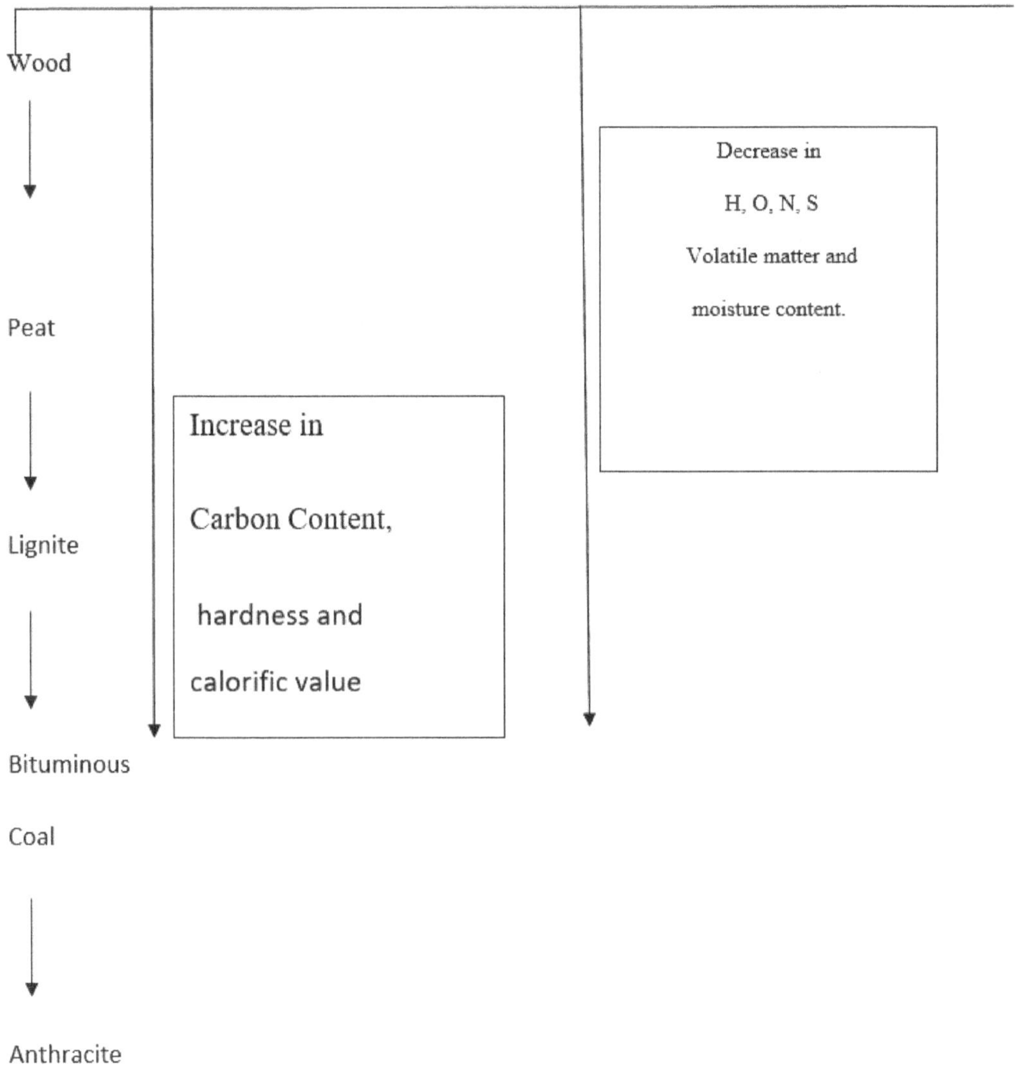

Peat

Peat is the first stage product of coalification process. It consists of partly decomposed plant material that has decomposed inside under marshy condition. Peat is brown, fibrous and jelly like in appearance. Peat is generally obtained by open cast mining. It contains 80 to 90 % of moisture and is considered an uneconomical fuel. After air drying for 1-2 months it burns freely. Its average composition is represented in abovetable . Calorific value of air -dried peat is around 5200 $Calg^{-1}$. In India peat deposits are located in Nilgiri hills (in Tamil Nadu).

Although not an economical fuel in India, it is used for power generation in England. Low temperature carbonization of peat produces peat coke a valuable fuel and for some metallurgical processes.

Lignite (Brown coal):

These are brown colored, soft and compact in texture and is considered as second stage coal after peat. Its moisture content is high but less than peat and breaks up easily into small pieces. Air dried lignite coal is dark brown in color and with lower moisture content. On drying lignite shrinks and breaks up readily in an irregular manner. It is used as house hold and boiler fuel as briquettes. However it is mostly used for manufacture of producer gas. In India Lignite deposits mainly occur at Neyveli (Tamil Nadu), Rajasthan, Kashmir and Assam.

Bituminous Coal (Common coal):

These are most common and widely used variety of coal. It is pitch black to dark grey in color. Its C content is around 90% with comparatively high calorific value. Depending upon its C content this common coal can be sub classified as follows:

i) **Sub-bituminous coal** – It is black, smooth with dull waxy lusture and homogeneous in appearance with C –content varies from 75-83%. It has lower moisture content of 12-25% and volatile matter content is very high and crumbles into small pieces on exposure to air and is therefore difficult to transport. They are also known as non-caking coal. Its calorific value is 7000 Cal g^{-1}.

ii) **Bituminous Coal** – It has a typical banded appearance with a C-content as high as 80-90%. This class of coal is used in industries for making metallurgical coke, coal gas, for steam raising and domestic heating.

iii) **Semi – bituminous coal** – This variety of bituminous coal is very rich in C-content (90-95%) with a low volatile matter. It has a calorific value of 8500 $Calg^{-1}$ and mainly used for manufacturing coke. In India Bihar, Bengal, Madhya Pradesh and Orissa have bituminous coal reserves.

Anthracite – It is the highest rank coal with 95% carbon content and lowest moisture and volatile matters. Anthracite coal is the hardest of all kinds of coal, quite dense and lustrous in appearance. It ignites without much difficulty, burns without smoke and provides intense local heating. They possess no caking power. It is used for raising steam, metallurgical purposes, household heating and places where smokeless high local heat is required. In India anthracite coal reserves are found in Kashmir and eastern Himalayas.

Cleaning of Coal

Coal preferably those with high ash content has to be cleaned to raise the efficiency of their utilization. Indian coals are usually of high ash content, cleaning of coal not only raises its efficiency of utilization but also lowers its contents of sulphur, phosphorous, chlorine, other harmful elements and other volatile matters. This cleaning of coal is also known as washing. Washing improves the caking properties of coal facilitating its transport and storage.

The minerals present in coal may be inherent (added during coalification process) or extraneous (added from external sources). Inherent mineral matter cannot be removed from coal by mechanical methods only. But extraneous mineral due to dirt bands, rock particles etc. which get mixed during mining, transportation or storage can be washed. Washing of coal is based on the principle of float and sink. Taking advantage of higher specific gravity of associated minerals than pure coal and by using liquids of different specific gravity, coal sample is divided into a number of fractions in the increasing order of specific gravity and hence the ash value.

Washing of coal

Coal mined from nature consists of variable sizes ranging from large blocks to fine dust. Again the directly mined coal is associated with various minerals such as slate and other impurities. Removal of these impurities and separation of coal according to sizes increases its calorific value.

The average specific gravity of coal is around 1.35 while that of impurities of bit higher (~2.0). Hence the calorific value of coal is improved by cleaning/washing coal of impurities using gravity separation or flotation method.

Gravity Washing

Mined coal is taken in a big tank known as coal washer, filled with water with dissolved calcium chloride. The whole mass is agitated violently by forcing water under pressure. Superior coal (being lighter) having specific gravity of about 1.6 forms the upper layer and taken out through the outlet. The lower layer contains mineral matter having high specific gravity is removed separately.

Floatation separation

In flotation process mined coal is taken in a solution of calcium chloride in water or a suspension of barium sulphate or clay in water and agitated strongly by passage of air. Pure coal floats while the heavier dirt and mineral matter settles down. The coal is separated and washed with water.

Jig washing

The mixture of mined coal and mineral matter is subjected to periodic upward and downward current of water. The upward flow disturbs the bed of coal and mineral matter while the downward flow facilitates the coal to get concentrated in the upper layer (being lighter). The concentrated coal is separated.

Storage of coal

Coal is very susceptible to aerial surface oxidation which reduces its caking property and calorific value. Also the heat generated during oxidation, sometimes sparks off combustion of coal heap in the storage area. Thus storage of coal needs special attention.

- Necessary ventilation must be provided in the storage area to dissipate the heat produced.

- Coal having high volatile matter is very susceptible to surface oxidation and can not be stored under water. Intermitant water spray on the coal heap must be provided so that its temperature should not exceed 90 ^{0}C for safe storage.
- However anthracite coal having low volatile matter can be stored under air.

Selection of Coal

The main parameter on which the quality of coal depends is its calorific value. However the selection of coal for different uses is also dependent on following factors.

1. Calorific value - A good coal should have high calorific value so that large quantities of heat can be obtained from a smaller amount of coal. This also reduces the cost of the process.
2. Moisture content – A good quality of coal should have low percentage of moisture to prevent heat loss.
3. Ash content – To enhance the heating value the ash content of coal should be low. It has been observed that a heat loss of 1.5% occurs due to presence of 1% of ash in coal.
4. Coking Quality – Coking coals are strong good quality coal. Coke is produced by high temperature Carbonization of coal.

$$\text{Coal} \xrightarrow{\text{Carbonization}} \text{Coke + Coal gas}$$

Depending on the nature of the product we get.

Caking coal – If product obtained is soft, plastic and fuse together to large coherent masses, then it is called caking coal. The original coal is called caking coal. This type of coal is difficult to be oxidized.

Coking coal – On the other hand if the product obtained is porous but hard and strong (coke), then the original coal is called coking coal. Being hard these types of coals are not crushed under heavy weight of ore and used in metallurgical purpose.

Non-Coking coal – The coals which on heating do not undergo any fusing effect are called non-coking coal or free burning coal.

Calorific Intensity – Calorific intensity may be defined as the maximum temperature reached when coal is completely burnt in the theoretical amount of air. Hence the calorific intensity of coal depends on

i) quantity ii) nature and iii) specific heat of the gaseous products of coal combustion. Thus

$$Flame\ temperature = \frac{Heat\ of\ combustion + Sensible\ heat\ of\ air}{\sum(Combustion\ products \times specific\ heat)}$$

Generally a fuel containing volatile matter burns with a flame and the total heat is liberated over a wide area and much less concentrated while fuels burning with out flames produce intense heat and the fuel is said to have high calorific intensity.

Size of coal – Uniform sizes of coal as a fuel favors easy handling, passage of air through it, mixing of gases and regulation of combustion. Thus coal with uniform sizes produces approximately uniform heating.

Sulphur and Phosphorous Content – A good quality of coal should contain low S and P contents. These undesirable substances called impurities are easily transferred from coke to metals and produce corrosive effects on metals. The oxides of S and P formed during combustion process on release to air pollute the environment.

Analysis of coal

The quality of coal is assessed by following two types of analysis.

i) Proximate Analysis

ii) Ultimate Analysis

Proximate Analysis – The analysis is named proximate as the data collected vary with the procedure adapted. It is an empirical but important analysis and deals with determination of moisture content, volatile matter, ash content and fixed carbon.

Moisture Content – The coal is always associated with some amount of moisture. The percentage of moisture present in coal is determined by thermal analysis. One gm of finely powdered and air dried coal is electrically heated at about 110 ^{0}C for 1 to 1.5 hours. Then it is cooled in a desiccator and weighted.

The percentage of moisture present in the coal is calculated from loss of weight of coal as follows:

$$\% \text{ of moisture content} = \frac{\text{Loss in weight}}{\text{Weight of coal taken}} \times 100$$

Volatile Matter

The above dried coal is weighted and further heated up to 930 ^{0}C in an electrical furnace for 10 minutes. At this temperature all the volatile matter present are driven out. Finally the coal is cooled and weighed. The weight loss in coal is due to removal of volatile matter.

$$\textbf{\%Volatile matter} = \frac{\text{Loss of weight}}{\text{Weight of coal taken}} \times 100$$

Ash Content – The above coal (after removing volatile matter) is weighed and further heated for half an hour at 750 ^{0}C. It is cooled and weighed. The process of heating, cooling and weighing is repeated till a constant weight is obtained. The ash content of the coal is calculated as follows:

$$\% \text{ of Ash} = \frac{\text{Loss of weight}}{\text{weight of coal taken}} \times 100$$

Fixed Carbon:

The amount of carbon that remains intact in the coal after combustion (i.e. not available for combustion) is known as fixed carbon of coal. Higher the fixed carbon content of the coal, the higher will be the calorific value of it. It is calculated as follows:

% of fixed carbon = $100 - \%(moisture + volatile content + ash)$

2) Ultimate Analysis

Coal is subjected to elemental analysis to determine its C, H, N_2, S, P and O_2 content which is otherwise known as ultimate Analysis.

i) Carbon and Hydrogen – For determining the C and H content about 1 to 2 gm of accurately weighed coal sample is subjected to combustion in a current of oxygen. The C and H present in the coal are converted to CO_2 and H_2O respectively. These are absorbed in previously weighed KOH and $CaCl_2$ tubes where the salts are formed as follows:

$$C + O_2 \rightarrow CO_2$$

$$2KOH + CO_2 \rightarrow K_2CO_3 + H_2O$$

And

$$H_2 + \frac{1}{2}O_2 \rightarrow H_2O$$

$$CaCl_2 + 7H_2O \rightarrow CaCl_2.H_2O$$

From the increased weight of KOH and $CaCl_2$ tubes the percentage of C and H content of the coal is calculated as follows:

$$\text{Percentage of C} = \frac{\text{Increase in weight of KOH} \times 12}{\text{Weight of coal taken} \times 44} \times 100$$

And

$$\text{Percentage of H} = \frac{\text{Increase in weight of CaCl}_2 \times 2}{\text{Weight of coal taken} \times 18} \times 100$$

ii) **Nitrogen:** Nitrogen being an inert and incombustible gas, its presence decreases the calorific value of coal. Thus a good Nitrogen content of coal is estimated using Kjeldahal's method. About 1 gm of accurately weighed coal is heated in a Kjeldahal's flask with concentrated H_2SO_4 and K_2SO_4 (catalyst) till a clear solution is obtained. This is followed by addition of KOH in excess to liberate ammonia. The liberated ammonia is absorbed in a known volume of a standard acid solution. The unused acid is then determined by back titration with standard NaOH solution. From volume of acid used by liberated ammonia, the percentage of N in coal is estimated as follows:

Percentage of N = $\frac{Volume\ of\ acid\ used\ into\ normality}{Weight\ of\ coal\ taken} \times 1.4$

iii) **Sulphur-** For determining S content of coal its washings from Bomb calorimeter (used for determining calorific value) is used. The washings obtained from Bomb Calorimeter of known mass(1gm) of coal is taken. During the determination S is converted into Sulphate and are precipitated as by $BaSO_4$by treatment with $BaCl_2$.The precipitate is filtered, washed and heated to constant weight.

Percentage of Sulphur =

$$\frac{Weight\ of\ BaSO_4\ obtained \times 32}{Weight\ of\ coal\ taken\ in\ bomb\ calorimeter \times 233} \times 100$$

Elementary sulphide, pyritic and organic forms of S adversely affect the calorific value of coal. Moreover during carbonization, combustion and gasification it emits corrosive SO_2, H_2S and mercaptans that pollutes the environment. Since coke always retains considerable part of S in coal, metallurgical industries demand coal of low S content for coke making.

iv)**Phosphorous** – Phosphorous occurs in a very minute quantity in coal. Its determination is important when the coal to be used as coke in metallurgical processes.

v) **Oxygen-**The oxygen content of coal is usually determined by the method of difference as follows.

Percentage of oxygen = $100 - \%\ of\ (C + H + N + S)$

Although in this method errors of all other determination are reflected in the oxygen value but other direct methods is both laborious and time consuming. High oxygen content coals have high inherent moisture and low calorific value. Thus a good quality coal should have low oxygen content.

PULVERIZED COAL

The efficiency of a coal is improved through pulverization. Pulverized (powdered) coal has increased surface area which facilitates its combustion and provides higher calorific value. It also facilitates quick release of volatile matters present in it.

Most of the big installations like thermal power station, cement rotary kiln fire use pulverized coal in the burners for better efficiency. Coal is pulverized in a ball mill or impact mill. The degree of fineness of a coal depends upon its rank is measured by passing it through a 200 mesh of 0.77mm. The standard is up to 85% of pulverized anthracite coal should pass through the above mesh while it is 60% for brown and sub-bituminous coal. Not more than about 2% should be retained on a 50 mesh (0.3mm). Preheated air at 110-300 ^{0}C is passed over the coal during grinding. This raises the capacity of the pulveriser by drying the fuel. The air also serves to transport the pulverized coal to the burner and act as the primary air which constitutes about 15 to 20% of the total combustion air. The solid content of pulverized coal-primary air mixture is about 0.5 Kg/m^3 and the temperature at which it is fed to the burners is usually maintained at 70-80 ^{0}C.

Through pulverization low quality coal can also be used efficiently. The process provides an easy and judicious way to control the use of the natural resource.

METALLURGICAL COKE

Coke is preferred over coal in metallurgical processes due to following reasons:

i) Porosity of coke is better than coal helping in complete combustion of the coke.

ii) Coke contains less volatile matter than coal as during carbonization much of undesirable sulphur and other volatile matters present in coal are removed and coke burns with a short flame but with high calorific value.

iii) Coke is stronger than coal hence can withstand the pressure of over burden of ore, flux and fuel together in the furnace.

Carbonization of Coal

Coal is heated to a certain temperature to remove the volatile matter and moisture present and to make it hard, porous and thermally efficient for combustion. The process is called carbonization. Coal is carbonized to get coke. A coke is a hard porous carbonaceous material used mainly as a source of reactive carbon in blast furnaces, foundries and gasifiers. It is also used as a fuel in boilers and heating appliances

The process of carbonization is of three types namely (i) Low temperature carbonization at 600 ^{0}C. This process is used for production of semi coke or soft coke mainly used for domestic purpose
(ii) Medium temperature carbonization at 800 ^{0}C. This process have no commercial utilization but many commercial low temperature carbonization processes approach the medium temperature one
(iii) High temperature carbonization at 1000 ^{0}C is followed for production of hard coke and coal gas. Hard coke has great demand for metallurgical processes.

i) Low temperature carbonization

In this process coal is heated at 500 to 700 ^{0}C. Coke yield is about 77%. The coke obtained in this method is not mechanically strong having a calorific value about 6500 to 9500 Cal/kg. It contains 5 to 15% of volatile matter and burns easily giving a smokeless hot and radiant fire. The byproduct gas is also richer in heating value and used as gaseous fuel.

This coke is soft and mainly used as a domestic fuel.

ii) High Temperature carbonization

In this process coal is heated at 900-1200 ^{0}C. About 75% coke is obtained which is of better quality than obtained from low temperature method. The coke obtained is harder, stronger and more pure to be used in metallurgical processes. It contains very less quantity of volatile matter and moisture. It produces a smoky heating of high thermal efficiency with a calorific value of about 5500-6000 KCal/kg.

Table Characteristics of Low and high temperature Carbonization

Sl. No	Characteristics	Low temperature Carbonization	High temperature carbonization
1.	Heating temperature	500-700 ^{0}C	900-1200 ^{0}C
2.	Yield of coke	77%	75%
3.	Percentage of volatile matter in coke produced	5-15%	1-3%
4.	Mechanical strength of coke	Not mechanically strong	Good mechanical strength.
5.	Calorific value	6500-9500 Kcal/Kg	5500-6000 Kcal/Kg
6.	Hardness of coke produced	Soft	Hard
7.	Smoke produced on burning coal	Smokeless	Smoky
8.	Percentage of aromatic hydrocarbons in gas production.	Lower	Higher
9.	Use of Coke	Domestic	Suitable for Metallurgical purposes.

Manufacturing of Metallurgical Coke

There are two process followed for production of metallurgical coke.

i) Beehive coke oven

Coke was first manufactured in India towards the end of nineteenth century in Beehive Oven for the iron and steel works at Kulti. Being the cheapest process for manufacture of metallurgical coke about 5000 Beehive coke ovens are located in Jharia, Bokaro-Ramgarh, Karanpura and Raniganj coal field. The batteries contain 20 to 30 Beehive ovens each. By better control of combustion conditions the yield may be raised to 72% with decrease in time from 72 to 60 hours.

The oven is a dome shaped structure of height 2m and base of diameter 3.5m. It has a hole at the center of the roof and a door at one side. A number of ovens are built in a row with common walls between. It is made up of heat resistant clay bricks. About 5 to 7 tonnes of coal are charged in to the hot oven from the hole at the roof that forms a layer of 60-90 cm depth by a rake. Carbonization immediately starts and evolved gases burn off with the air coming through an opening at the top of the closed door. Carbonization continues for 2 to 3 days. Heat is transferred from roof to bottom then through layer of charge. Heat is supplied by burning of volatile matter and no byproducts are obtained. The exhaust gases are either allowed to escape to the atmosphere or passed through flues built in the walls between the oven and then released to the atmosphere. This increases the thermal efficiency and reduces the pollution. On completion of carbonization the door is opened and hot coke is quenched with water and raked out. The walls and roof of the oven are hot enough and initiates carbonization of next charge.

The main demerits of Beehive coke oven are:

i) Lack of flexibility in operation

ii) No by products are obtained

iii) Lower coke yield due to partial combustion

However the due to low running cost and production of extremely good quality strong of blocky coke of required reactivity by using only good quality coking coal are the advantages of Beehive coke oven.

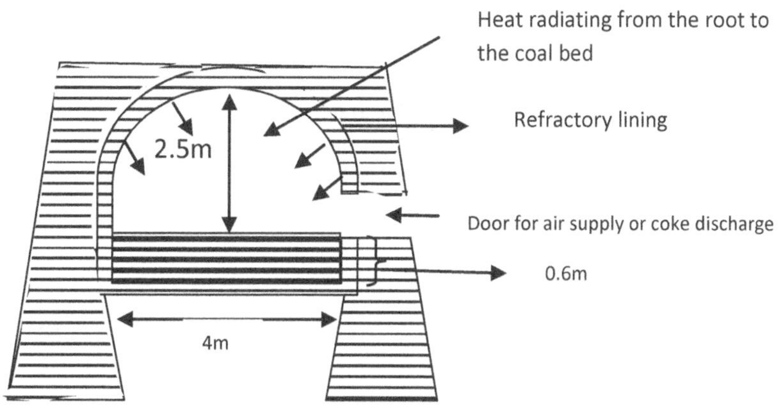

Fig : Beehive Coke oven

2. Otto Hoffman's by product Oven

This oven as developed by Otto Hoffman where coal is indirectly heated by combustion chambers separate from carbonization chamber. A gaseous fuel is burnt in the combustion chambers for supplying heat to the process. Coal is coked in the carbonization chambers. The gases evolved during process are treated for recovery of by products like tar, ammonia, crude benzol and coke oven gas.

The Otto Hoffman by product Oven is made up of number of narrow silica chambers (each about 10 to 12 m long and 3 to 4 m high with 0.40 to 0.45 m wide) erected side by side with vertical flues in between them to form a battery. The silica chambers are provided with a charging hole at the top and a gas off- take, a refractory lined cast iron door at each ends for discharging coke.

Finely divided coal (charge) is introduced through charging holes in to the silica chambers and heated at about 1200 ^{0}C without air by the gaseous fuels. On regenerative system of heat, the flue gases produced during combustion pass on their sensible heat to checker brick there by raising its temperature to about 1000 ^{0}C, enough to preheat the inlet gases. Thereafter the flue gases are chanellised to second checker bricks for heating it. The process is repeated continuously to utilise maximum gas sensible heat and evolution of volatile matter ceases. Carbonization of a charge of coal takes approximately 11 to 18 hours. On completion of carbonization the coke is taken out and quenched by water spray (wet quenching) or cooled by passing inert gases like nitrogen (dry quenching). Dry quenching is preferred to wet quenching as the coke obtained by the former method is more strong, dense, graphitized and non-reactive.

Recovery of by-products

The coke oven gas is rich in ammonia, H_2S, naphthalene, benzene, tar, moisture etc.

i) *Recovery of tar* – The gas is passed through a tower in which liquor ammonia is sprayed. The dust and tar gets collected here in a tank on further heating (by steam coils) ammonia is recovered for reuse.

Recovery of ammonia – The gases then passed through a tower where water is sprayed. Ammonia is collected here as NH4OH.

Recovery of naphthalene – The gases thereafter passed through another tower where water is again sprayed at low temperature thereby facilitating condensation of naphthalene.

Recovery of benzene – The gases are then sprayed with petroleum, when benzene and its homologues gets separate

Liquid fuels

Petroleum – Petroleum or crude oil is the source of many liquid fuels of present day. The name petroleum originates from the Latin word Petra means rock and oleum is oil. Thus, petroleum or rock oil is a dark greenish brown viscous liquid found deep in earth crust. For the reason petroleum is also known as crude oil or mineral oil. This oil is usually found floating upon a layer of brine and accompanied by gases called natural gases.

COMPOSITION:

The composition of crude petroleum varies from field to field, however the average composition of petroleum can be expressed as:

C = 79.5 to 87%	S = 0.1 to 3.5%
H = 12 to 15%	N = 0.1
	O = 2 to 3 %

The ash of crudes is only 0.1 % or less. Practically all metals have been found in petroleum ash, the most widely occurring elements are Si,Fe,Al,Ca,Mg,Ni and Na., Crude petroleum is a mixture of hydrocarbons like paraffins, cycloparaffins (naphthalene) and aromatics in varied proportions. Olefins and acetylene are absent in crude oil. Olefins formed during processing of petroleum.

Classification:

The composition of crude petroleum varies widely from field to field so also its chemical properties. However, the three-group classification system of crude petroleum used by petroleum refiners for commercial practices are as follows:

Paraffin-base-crude : Mainly contains saturated hydrocarbons up to $C_{35}H_{72}$ and a little of aromatics including napthalene. It produces a residue containing both wax and asphalt.

Asphalt base crude – Mainly contains cycloparaffins (naphthalene) with smaller amount of paraffins and produces black lustrous residue of asphalt. Due to presence of large amount of naphthamic hydrocarbons. It is also known as naphthene base crude.

Mixed base crude – Contains both paraffinic and asphaltic hydrocarbons and produces a residue containing both wax and asphalt.

Table No. Comparison of paraffin-base and Asphalt base crudes

Properties	Paraffin base	Asphalt base
Specific gravity	Low	High
Yield of straight run gasoline	High	Low
Octane number of gasoline	Low	High
Odour of gasoline	Sweet or sour	Aromatic or sour
Sulphur content of fractions	Low	High
Smoke point of kerosene	High	`Low
Cetane number of diesel	High	Low
Pour point of diesel	High	Low
Yield of lubricants	High	Low
Viscosity index of lubricants	High	Low

Origin of Petroleum

A large number of theories have been suggested regarding genesis of petroleum.

Organic or Engler's Theory

According to this theory marine animals, fish and other aquatic vegetable matter like sea weeds etc. are covered up during volcanic eruptions, could have undergone decomposition in dead-block sea area under the influence of high pressure and temperature. This theory is able to account for the presence of brine and fossils in petroleum. However it fails to tracc thc origin of chlorophyll and radioactive compounds in petroleum.

2. **Carbide theory (Mendeleev's theory)**

Carbide theory predicts the origin of petroleum from inorganic source. According to this theory molten metals like Fe, Al, Ca etc present in hot interior of earth reacted with coal deposits under pressure and high temperature to form carbides which led to the formation of hydrocarbons on hydrolysis.

$$Ca + 2C \xrightarrow{hightemp\&pressure} CaC_2$$

$$CaC_2 + 2H_2O \longrightarrow Ca(OH)_2 + C_2H_2$$

$$4Al + 3C \underset{andpressure}{\overset{Hightemperature}{\rightarrow}} Al_4C_3$$

$$Al_4C_3 + 12H_2O \rightarrow 4Al(OH)_3 + 3CH_4$$

These lower hydrocarbons then give rise to paraffins and aromatic hydrocarbons on hydrogenation and or polymerization

$$C_2H_2 \overset{H_2}{\rightarrow} C_2H_4 \overset{H_2}{\rightarrow} C_2H_6$$

$$3C_2H_2 \rightarrow C_2H_4 \overset{H_2}{\rightarrow} C_2H_6$$

$$3C_2H_2 \rightarrow C_6H_{12}$$

$$2C_2H_4 \rightarrow CH_3CH = CHCH_3 \overset{H_2}{\rightarrow} CH_3CH_2CH_2CH_3$$

Butane

However the theory is unable to account for the presence of N ,S Chlorophyll, haemin, brine, optically activity of petroleum organics.

Modern Theory

According to the modern theory origin of petroleum is due to partial decomposition of marine animal and vegetable organisms of prehistoric forests. Thus transformation was influenced by

i) Presence of radioactive substances like Uranium or by ii) bacteria influenced anaerobic decomposition in presence of clay minerals, strongly reducing condition, temperature and high pressure. Presence of organic compounds namely phytane, chloestane and porphyrins in petroleum corresponds to natural compound like phytanic acid, chloesterol, chlorophyll favors the organic origin for petroleum.

Mining of Petroleum

Generally oil is found in porous strata beneath the impervious rock at a depth varying from 500 to 15000 feet. Mining of oil is done by drilling holes in earths crust and inserting pipes upto oil bearing porous rock. Initially as the hydrostatic pressure of the natural gas present in oil bed being very high, oil gushes out through these pipes by itself. Later on as the pressure decreases the oil is mechanically pumped out using lift pumps or air lift pumps.

The air lift pump comprises of two coaxial pipes, lowered to the base of the oil bed. Air is blown through the outer pipe which results in pushing out oil through the inner pipe. The crude oil thus collected through mining is conveyed to refinery for refining. India is having 42% of its total area as oil yielding sedimentary rocks while in offshore the sedimentary basin extends upto a depth of 200 ms.

Indian Resources of Petroleum

India's first oil well was drilled in Assam near an oil seepage area namely Nahar Pong in 1866. The oil well was drilled upto 31m deep but failed to strike oil. In the following year 1867 oil was discovered. Assam in an oil well at 35.5 m at Makum in Digboi area of Assam. More and more oil wells were dug to initiate commercial product of crude oil. Nearly nine million tonnes of crude oil have been produced from Digboi oilfield in about a thousand oil wells. Initially crude oil was refined in the small refinery at Margherita. In 1899 the first full fledged refinery was commissioned at Digboi with a capacity of 90 kl per day.

Through the exploration work of government and semi-government agencies several more oil fields have been discovered namely – Ankleswar, Kalol, Sanand, Nawagam, North Kadi, Hugrijan, Moren-Nahorkatiya, Rudrasagar, Lakwas, Galkei, Bombay high and cambay in the western offshore, Palk strait in south and the Godavari off- shore in the east. According to the statistics India produces 49% of crude from on shore area while the remaining 51% is from off shore oilfields. At present India is having 18 refineries with a total capacity of 127.37 Mt .Four more refineries are to come up at Paradip (orissa). According to the report of the Ministry of petroleum and natural gas 2013.

Table Location and capacity of functioning refineries in India.

Sl No.	Name of the company	Location of the refinary	Capacity million metric ton per annum
1.	Indian oil corporation Ltd. (IOCL)	Guwahati	1.0
2.	IOCL	Barauni	6.0
3.	IOCL	Koyali	13.7
4.	IOCL	Haldia	6.0
5.	IOCL	Mathura	8.0
6.	IOCL	Digboi	0.65
7.	IOCL	Panipat	6.0
8.	Hindustan Petroleum	Mumbai	5.5
9.	HPCL	Visakhapatnam	7.5
10.	Bharat petroleum corporation Ltd. (BPCL)	Mumbai	6.9
11.	Chennai Petroleum corporation Ltd. (CPCL)	Manali	9.5
12.	CPCL	Nagapattnam	1.0
13.	Kochi Refinery and petrochemicals Ltd. (BRPL)	Kochi	7.5
14.	Bongaigaon refinery and petrochemicals Ltd. (BRPL)	Bongaigaon	2.35
15.	Numaligarh Refinery Ltd (NRL)	Numaligarh	3.0
16.	Mangalore Refinery and petrochemicals Ltd. (MRPL)	Mangalore	9.69
17.	Tatipaka Refinery (ONGC)	Andhra Pradesh	0.078
18.	Reliance Petroleum Ltd. (RPL)	Jamnagar	33.0

Nature of Indian Crudes

The oil fields of India yield highly waxy to mixed base crudes. The crude obtained from Bombay High is highly aromatic almost sulphur free with lowest wax content and is the lightest while cambay crude is the heaviest as the wax content being highest. However the sulphur content is very low in Indian crudes.

Refining of crude oil

The crude oil is an extremely stable suspension of oil and brine (salt water) along with minor quantity of sand. After removing water and other objectionable (from example S) material the crude oil is separated into various fractions through fractional distillation process. The process of separation of crude oil into different useful petroleum products is called refining of crude oil which is done in specially designed plants called refineries.

Refinery processes may be conveniently grouped as

1. physical separation process like distillation, solvent extraction etc.

2. Breakdown process like cracking visbreaking and coking.

3. Rebuilding processes – like reforming, alkylation, isomerisation, polymerisation etc.

1. Removal of water

Cottrell's Process

The crude oil which is a stable emulsion of oil and brine is allowed to flow between two highly charged electrodes. The colloidal water droplets coalesce (aggregates) to form large drops and separates out from the oil.

Persistent impurities, colouretc are removed by passage of the crude oil over adsorbents like Kieselguhr, clay etc.

2. Removal of harmful sulphur compounds

Crude oil obtained from different oil fields of India possesses a very low percentage of S. To remove Sulphur, the crude oil is treated with copper oxide which removes the sulphur present in the crude as copper sulphide precipitate. The crude oil is freed from solid copper sulphide through filtration.

3. Fractional Distillation

The crude oil is heated in an iron retort to a temperature of about 400 ^{0}C in a furnace. All the volatile constituents are evaporated leaving the asphalt or coke as a residue. The hot vapors are then passed up a "fractionating column", which is a tall cylindrical tower containing a number of horizontal stainless steel trays at short distances. Each tray is equipped with a small chimney which is covered with a loose cap. As the vapours go up they get gradually cooled and condense at different heights of the column. The higher boiling fractions will condense first followed by condensation of lower boiling fractions. The lower boiling fractions can be purified and used directly while the higher boiling fractions get collected at the bottom are subjected to 'cracking to more useful low boiling fractions. The various fractions of products obtained from fractional distillation of crude oil is listed below in Table

Table: Fractional products from crude oil

Sl. No.	Fraction	Boiling Range in ^0C	Approximate composition	Applications
1.	Uncondensed gas (Refinery gas)	Upto 30 ^0C	C_1 to C_4	Domestic and Industrial fuel (LPG)
2.	Petroleum Ether	$30^0 – 70^0$	C_5 to C_7	Solvent for fat, oil, varnishes and rubber etc
3.	Gasoline, Petrol or Motor spirit	$40^0 – 120^0$	C_8 to C_9	Motor fuel, solvent and dry cleaning
4.	Naptha (solvent spirit)	120^0 - 180^0	C_9 to C_{10}	Solvent and dry cleaning
5.	Kerosene Oil	$180^0 – 250^0$	C_{10} to C_{16}	Fuel,illuminetant,Oil gas
6.	Diesel oil or fuel oil or gas oil.	250^0-320^0	C_{10}-C_{18}	Diesel engine fuel
7.	Heavy oil	$320^0 – 400^0$	C_{17}-C_{30}	For getting gasoline by cracking.
8.	Lubricating oil (refractionated heavy oil)		C_7-C_{26}	As lubricant
9.	Vaseline			As lubricant
10.	Grease			As lubricant
11.	Paraffin wax			Candles, boot polish, tarpaulin etc.
12.	Residue a) Asphalt (petroleum coke).	Aove 400^0	C_{30} and above	Road topping water-proffing of roofs as a fuel in moulding work.

Amongst the above products the fractions widely used as liquid fuel is as follows:

i) Gasoline or petrol fraction : This product fraction is obtained between the range 40^0 to 120^0 C. Gasoline is a mixture of hydrocarbons namely pentanes (C_5H_{12}) to octanes (C_8H_{18}). Gasoline is highly volatile and inflammable with a calorific value of 11,250 Calg^{-1}. It is used as a fuel for internal combustion engines of automobiles and aeroplane.

ii) Diesel Oil – It is obtained a product fraction between the range of temperature 250 ^{0}C- 320 ^{0}C. It is a mixture of higher hydrocarbons (C_{15} to C_{18}). The calorific value of diesel oil is 11000 Calg^{-1} and is used as a diesel engine fuel.

iii) Kerosene Oil – It is obtained between the temperature range of 180 – 250 ^{0}C and is a mixture of hydrocarbons from decanes ($C_{10}H_{22}$) to hexadecanes ($C_{16}H_{34}$). It does not vaporise easily and has a calorific value of diesel oil is 11,100 Calg^{-1}. Due to its high boiling point range kerosene oil is widely used as domestic fuel in stoves as jet engine fuel and for making oil gas.

CRACKING

The process of decomposition of less volatile (high boiling point) higher molecular weight hydrocarbons into more volatile (low boiling point) lower molecular weight hydrocarbons by application of heat with or without catalyst is known as cracking.

$$C_{10}H_{22} \xrightarrow{Cracdking} \underline{C_5H_{12} + C_5H_{10}}$$

Decane Pentane Pentene

Boiling point = 174 ^{0}C, Boiling point 36^0 to 150 ^{0}C

$$C_{12}H_{26} \xrightarrow{Cracking} C_7H_{16} + C_5H_{10}$$

Dodecane B.Pt. 30^0 – 150 ^{0}C

Bpt = 216 ^{0}C

The quality of gasoline obtained directly from fractional distillation of petroleum (straight run gasoline) is not high and needs proper blending. Moreover compare to the demand of gasoline as a motor fuel mainly the yield is very less about 16-26 % only. To increase the yield the surplus middle and heavy fractions are subjected to cracking process. Cracking not only increases the yield of petrol but also improves the quality of gasoline in terms of antiknock properties than straight run gasoline.

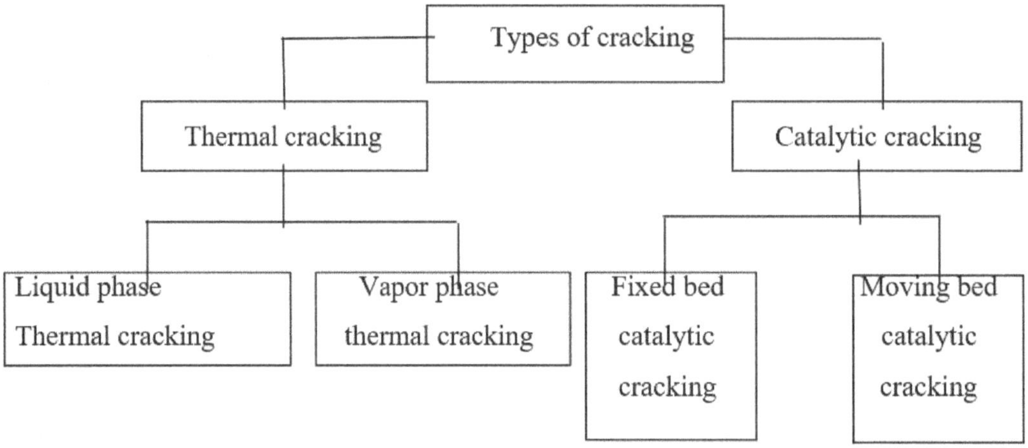

Thermal cracking

The thermal cracking of heavy oils are carried out at high temperature and pressure without any catalyst. High temperature (greater than its boiling point) facilitates the rupture of carbon carbon bonds while high pressure prevents vaporisation of the cracked residue. In thermal cracking the feed is heated upto 450 ^{0}C to 750 ^{0}C at pressures ranging from 1 – 70 atm. The important reactions are decomposition, dehydrogenation, isomerism and polymerisation as follows:

Decomposition of paraffinic molecule to low molecular weight hydrocarbon

$$CH_3(CH_2)_8CH_3 \rightarrow CH_3(CH_2)_4CH_3 + CH_3CH_2CH = CH_2$$

n-decane n-hexne butane-1

while n-decane is a component of gas oil, n-hexane is one of the gasoline. The low molecular weight product obtained then undergoes isomerisation, dehydrogenation, polymerization reactions as follows

H CH3 CH3

| | |

$2CH_3CH_2 - CH = CH_2 \rightarrow CH_3$ - C – C - C = CH$_2$

| |

H H

Butene – 1 Iso-Octane

Iso-octane is a component of gasoline. Along with the liquid paraffin, gases and solids (coke) are also produced during cracking. The ease of cracking depends upon the nature of the feed and its boiling range. For example heavier fractions and paraffinic hydrocarbon have higher crackability than low molecular weight and aromatic one respectively. Depending on the feed, thermal cracking is done either in liquid phase or in vapour phase.

Liquid phase thermal cracking:

The paraffinic feed is heated at 475 – 530 ^{0}C at a pressure range of 7-70 atm to keep the reaction product in the liquid state. The cracked products are separated in a fractionaling column. The yield of petrol is 60-65% and the octane rating is 60-70.

Vapour phase thermal Cracking

The liquid paraffinic feed in first vapourised by heating at about 600 ^{0}C at a pressure of 4-10 atm. This process is used for highly volatile oils. The yield is about 70%. The petrol produced has better antiknock properties but lower in stability than that liquid phase cracking.

Catalytic Cracking

The yield and quality of gasoline from cracking can be greatly improved by use of a suitable catalyst like aluminium silicate $[Al_2(SiO_3)_3]$and alumina $[Al_2O_3]$or acid treated natural clays of montmorillonite group.Recently zeolite is also used as a cracking catalyst with better activity. Catalysts are used as pellets beads or powder form.All new cracking units for gasoline production are based on catalytic cracking process. Thermal cracking units are now restricted to olefinic production, visbreaking (reduction of viscosity of straight run products) and coking.

Merits of Catalytic Cracking Process

1. Better quality petrol is produced.
2. Yield of petrol is higher
3. No external fuel is required for cracking as the heat required for cracfking is derived from the coal embedded in the catalyst.
4. A much lower pressure (~1-2 atm) is needed thereby reducing the production cost.
5. Catalytic cracking process can be easily controlled and hence ensures formation of the desired.
6. Higher aromatic content improves the antiknocking property of petrol.
7. S content of product (petrol) is very low.
8. Selective action of catalyst facilitates cracking of only high boiling fraction of paraffins.

Catalytic cracking of paraffins are done by following methods:

a) Fixed bed catalytic cracking

The heavy oil charge is heated to cracking temperature (420-450 ^{0}C) in a pre heater and then passed through a catalytic chamber (containing either silica-aluminium gel $SiO_2 - Al_2O_3$or bauxite or artificial clay mixed with zirconium oxide as catalyst) maintained at 425-450 ^{0}C at 1.5 Kg/Cm^2 pressure.

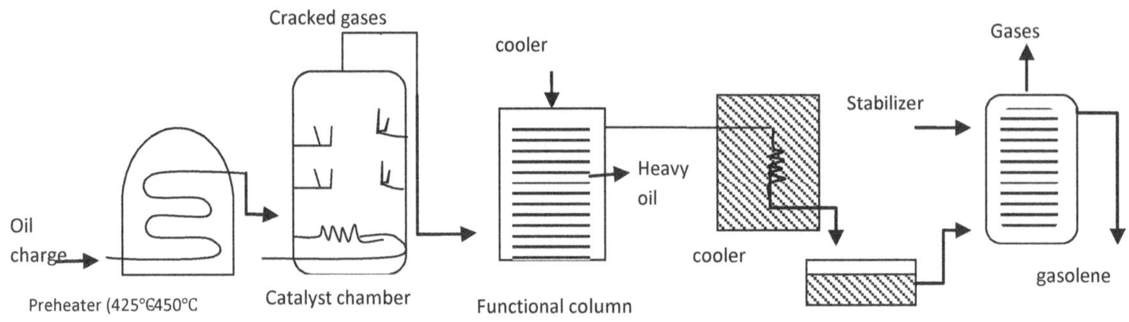

Figure : Fixed bed Catalytic cracking

Cracking of the heavy oil vapours takes place in the catalytic chamber. About 40% of the charge is converted to low molecular weight hydrocarbons. About 2-4% of carbon which is also formed during cracking process gets adsorbed on the catalyst and deactivates it. The cracked vapours are passed through the fractionating column,where gasolinevapours condense along with some other gases and the uncondensed gases move up. The gasoline is than passed through the stabilising column where the dissolved gases are removed from the top and pure gasoline is recovered.

During the process the catalyst stops functioning after about 10-11 hours due to deposition of carbon particles on it formed in the process. This is reactivated by burning off the carbon in a steam of hot air. During the reactivation process the vapours of heavy oil are diverted through another stand of catalytic chamber and the continuity of the process is maintained.

The cracking being an endothermic process the temperature is the catalytic chamber is maintained by circulating a mixture of molten sodium nitrate and sodium nitrite in pipes through the catalytic bed.

b) **<u>Moving Bed (Fluidised Bed) Catalytic Cracking:</u>**

The solid catalyst is very finely powdered to be maintained as a free flowinglosely packed suspension by feed stock vapours. This finely powdered catalyst get mixed with heavy oil vapours and behave much like a fluid hence the name " fluidised bed."

The vapours of heavy oil mixed with hot catalyst (fine powder) is forced up into a large reactor 'bed' maintained at 500 ^{0}C. Although the high temperature of the catalyst (600 $_0$C) initiates the cracking process of heavy oil much before its entry into the reactor and facilitates the cracking of high boiling fractions. After cracking of the heavy oil, cracked vapours are passed into the fractionating column through the internal cyclone, fixed at the top. The internal cyclone (a centrifuge)(allows only the cracked vapours to pass through it and retains the catalyst powder in reactor shelf. The catalyst powder gradually becomes heavier due to carbon deposition and settles at the bottom, from where it is forced by an air blast to the regenerator (maintained at 600 ^{0}C). In regenerator carbon is burnt off and the regenerated catalyst then flows through a stand pipe for mixing with a fresh batch of cracking oil. The cyclone fixed at the top of the regenerator allows only gases like CO_2 to pass out retaining the catalyst particles. The heat liberated in the regeneration process is used to preheat the catalyst.

Mechanism

The mechanism for cracking of molecular weight hydrocarbon to low molecular weight depends on the process followed.

i) Thermal cracking : Thermal cracking of high molecular weight hydrocarbons follows free radical mechanism. The mechanism has been illustrated with n-nonane.

Initiation

Free radicals are produced when n-nonane undergoes homolytic cleavage by the high temperature of cracking

$$CH_2(CH_2)_7CH_3 \xrightarrow{Thermal} CH_3(CH_2)_3CH_2 + CH_3(CH_2)_2CH_3$$

Propagation

The free radicals generated in the initiation step undergo fission at β-position to yield a new radical and an olefin

$$CH_3-(CH_2)_3-CH*_2 \rightarrow CH_3CH_2CH^*{}_2 + CH_2 = CH_2$$

$$CH_3-CH_2-CH^*{}_2 \rightarrow CH^*{}_3 + CH_2 = CH_2$$

The propagation reaction continues till a thermally stabler radical is formed.

The radical having no C-C in the β-position generated hydrogen free radical as follows:

$$CH_3-CH^{\bullet}{}_2 \rightarrow CH_2 = CH_2 + H^{\bullet}$$

Termination

Termination process takes place as follows:

$$\dot{C}H_3 + \bar{C}H_3 \rightarrow CH_3-CH_3$$

$$CH_3-\dot{C}H_2 + \dot{C}H_2-CH_3 \rightarrow CH_3-CH_2-CH_2-CH_3$$

Or $CH_3\dot{C}H_2 + \dot{H} \rightarrow CH_3-CH_3$

b) Catalytic cracking

Catalytic cracking of hydrocarbon belived to involve carbonium-ion intermediates. First the high temperature induces thermal cracking producing a small amount of alkene which is thereafter protonated by the catalyst.

$$C_9H_{20} \xrightarrow{Thermal} CH_3-CH_2-CH = CH_2 + C_5H_{12}$$

$$CH_3-CH_2-CH = CH_2 + H^+ \rightarrow CH_3-CH_2-\overset{+}{C}H-CH_3$$

$$CH_3-(CH_2)_3-\underset{\displaystyle H}{\underset{|}{CH}}-(CH_2)_3-CH_3 \rightarrow CH_3-(CH_2)_3-\overset{+}{C}H-(CH_2)_3-CH_3$$

$$+CH_3-CH_2-CH_2-CH_3$$

The carbonium ion undergoes β −cleavage to produce an alkene and a smaller carbonium ion.

$$CH_3-CH_2-CH_2-CH_2-CH-CH_2-CH_2-CH_2-CH_3$$

$$\rightarrow CH_3-CH_2-CH_2+CH_3-(CH_2)_3-CH=CH_2$$

The primary carbonium ion rearranges to the more stable secondary carbonium ion by 1,2 hydride shift.

$$CH_3-CH_2-\overset{\oplus}{C}H_2 \rightarrow CH_3-\overset{\oplus}{C}H-CH_3$$

with aromatic hydrocarbon, the rings remain intact only the side chains are affected

Synthetic Petrol

Excluding from natural resources, petrol can also be synthesized from cheaper sources as follows.

i) **Polymerization** – The gases obtained as a byproduct during cracking of heavy oil contains low molecudlar weight olefins and alkanes like ethene, propene, butene, methane, ethane, propane and butane respectively. This gaseous mixture when subjected to high temperature and pressure with or without catalyst undergoes polymerization to yield higher hydrocarbons resembling gasoline called polymer gasoline.

$$CH_3-CH=CH_2+CH_2CH_2CH=CH2 \xrightarrow[\text{High pressure high temperature}]{\frac{and}{or}\ catalyst} CH_2=CH-CH_2-CH(CH_3)_2$$

5-methyl hexene

The polymer gasoline thus obtained possesses high octane number (78+83)

On passing cracked gases through 70% H_2SO_4at 10-20 ^{0}C, isobutylene dimerizes to di-iso butylene leaving behind other olefins unaffected.

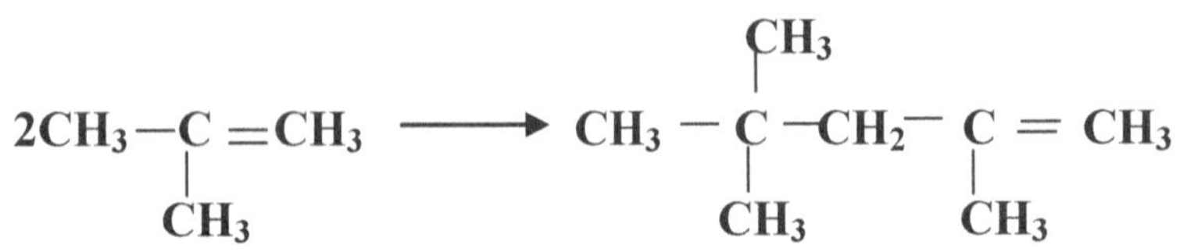

Isobutene Di-isobutylene (iso-Octane)

The di-isobutylene thus obtained is hydrogenated in presence of Ni as a catalyst at 160 ^{0}C to yield iso-octane.

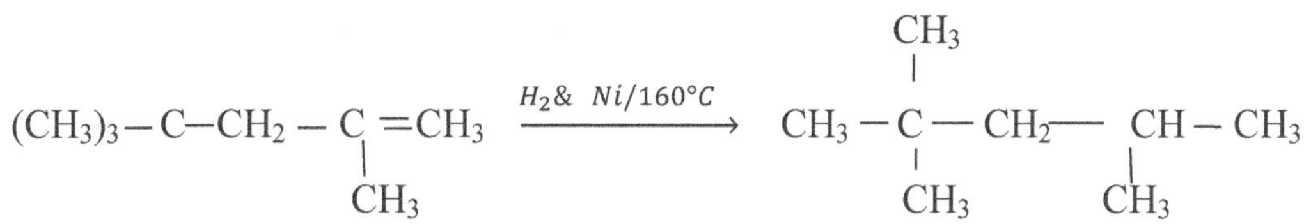

Iso Octane

Iso-octane has octane number of 100 polymerization reaction can be carried out in following two ways.

i) Thermal polymerization

In this method the cracked gases are subjected to a temperature of 500-600 ^{0}C and pressure of 70-350 Kg/Cm^2. The products are gasoline and gas oil mixture, which are further subjected to fractional distillation for separation.

ii) Catalytic polymerization

In this method polymerization of cracked gases are carried out in presence of a catalyst like phosphoric acid, sulphuric acid etc at a lower temperature (150-200 ^{0}C). The products formed are gasoline and unpolymerized gas. The later is separated and recycled for polymerization.

2. **Fischer-TropschMethod** : The method is based on synthesis of gasoline from hydrogenation of water gas, $(CO + H_2)$.

$$nCO + (2n+1)H_2 \rightarrow C_nH_{2n+2} + nH_2O$$

$$nCO + 2nH_2 \rightarrow C_nH_{2n} + nH_2O$$

The method was developed by Fischer and Tropsch in 1962 in Germany. Water gas required for the purpose is obtained by passing steam over red hot coke (obtained from coal).

$$C + H_2O \rightarrow CO + H_2$$

Coke steam water gas

The water gas is purified from H_2S by passing through Fe_2O_3 and thereafter through a mixture of Fe_2O_3 and Na_2CO_3 to remove organic sulphur compounds present. The purified gas is compressed to a 5 to 25 atm pressure and led into a catalytic converter. The catalyst consists of a mixture of cobalt (100 parts), thoria (5 parts) magnesia (8 parts) and Kiselguhar (200 parts), maintained at 200-300 ^{0}C. A mixture of saturated and unsaturated hydrocarbons result.

The reaction being exothermic, the gaseous mixture coming out of the reaction chamber are passed through a cooler where resembling crude oil is obtained. The crude oil on fractional distillation yields gasoline, diesel and heavy oil. The heavy oil is reused for cracking to produce gasoline.

Gasoline produced in Fischer-Tropsch method is rich in straight chain hydrocarbons, hence octane number is only 40. However the diesel oil is of excellent quality.

3) **Bergius Process** : This method of synthetic gasoline production was developed by Bergius. In this process low grade coals like bituminous or brown coals are converted into liquid and gaseous fuel by catalytic hydrogenation.

Fine powder of low rank coal is mixed with heavy oil and a catalyst (composed of tin, nickle oleate or molybdenum) to form a paste. The paste formed is heated in a currrent of hydrogen at 450-490 ^{0}C and 200 atm pressure for about 2 hours. During this period hydrogen combines with coal to form saturated hydrocarbons, which decomposes at the reaction temperature and pressure to yield low boiling liquid hydrocarbon. The vapours produced in the catalytic converter are led through a cooler where a liquid resembling crude oil is obtained. On fractionation this crude oil produces gasoline, middle oil and heavy oil respectively. The heavy oil is used to produce paste with fresh coal dust while the middle oil is hydrogenated in vapour phase in presence of a solid catalyst to yield more gasoline. The yield of gasoline in this method is 60% of the coal dust used.

Bergius method of synthetic petrol production although not economical yet practiced in countries which have no oil resources but rich in coal resource.

Refining of Gasoline:

The gasoline obtained from crude oil or by synthetic process contains undesirable straight chain alkenes and sulphur compounds. The alkenes can undergo oxidation and polymerization leading to **gum and sludge formation on storage, while S compounds (mercaptans) present can be instrumental** for corrosion of internal combustion engines besides producing undesireableodour in gasoline.

Removal of these undesirable substance is performed as explained below.

i) **Removal of Sulphur Compounds:**

Mercaptan containing gasoline is called **'sour' gasoline**, while (The sulphur compounds are removed by treating sour gasoline with sodium plembite with controlled addition of sulphur

$$Na_2PbO_2 + 2RSH \rightarrow Pb(SR)_2 + NaOH$$

(Doctor solution) Led Mercaptide

$$Pb(SR)_2 + S \rightarrow PbS_{(S)} + RSSR$$

The Pbs formed is removed through filteration while the disculphide being oil soluble is extracted. This refining process is termed as '**sweet petrol'** and solution (alkaline sodium plumbite) used for refining as '**Doctor Solution'**

ii) **Removal of Olefins** – The unsaturated hydrocarbons or olefins are removed by treating the gasoline with 80% H_2SO_4 . This does not react with paraffins, naphthalenes and aromatics. Only unsaturated hydrocarbons react with 80% H_2SO_4 forming alcohol and esters followed by polymerizing to tar like substances which are later removed by filteration.

iii) Further the gasoline is purified by percolating through fuller's earth which preferentially removes any olefins left and colours through adsorption.

Some inhibitors are also added to the refined gasoline to suppress any further oxidation reactions in it. This also improves the storing quality of gasoline.

TABLE: Comparative study of gasoline.

Sl. No	Straight-run gasoline	Cracked gasoline	Polymer gasoline
1.	It is obtained from straight distillation of crude petroleum	It is obtained from heavy oil.	It is synthesized from low molecular weight components.
2.	It is physically separated from crude oil.	It is a chemical process in which high molecular weight hydrocarbons are broken down to low molecular weight hydrocarbons.	It is a chemical process in which low molecular weight hydrocarbons polymerizes to yield high molecular weight hydrocarbons.
3.	It has low octane number	Higher octane number than straight run gasoline.	It is having highest octane number.
4.	It contains normal alkanes	It contains more aromatics, naphathenes and branched chain hydrocarbons.	It contains more branched chain hydrocarbons.
5.	Composition varies from C_5 C_9	Composition range is narrow from C_6-C_8	Composition range is limited to C_7 to C_8.

Knocking

Knocking can be described as a sharp metallic sound like rattling of a hammer produced in internal combustion engines, due to immature ignition of air-gasoline mixture.

A mixture of gasoline or diesel with air is used as a fuel in internal combustion engines, where the mixture of fuel (gasoline or diesel with air) is ignited in a cylinder. For petrol engines the ignition is brought about by means of an electric spark while in diesel engines it is done by compressing.

Petrol Engine Fuel

The whole cycle of four stroke in a petrol engine can be summarised as follows:

i) Suction stroke – The fuel vapors along with air present in the carburetter is drawn into the cylinder during induction process.

ii) Compression stroke – In this step the fuel air mixture is compressed.

iii) Power stroke – The mixture is ignited by an electric spark and the hot gases produced due to combustion, increases the pressure which pushes the piston down the cylinder.

iv) Exhaust Stroke – Finally the piston ascends and expels the exhaust gases from the cylinder and the next cycle follows.

For an efficient internal combustion engine:

The fuel mixture must burn smoothly and rapidly which is possible after initiation of combustion by a spark the flame spread rapidly and smoothly through the gaseous mixture. Thereby expanding gas drives the piston down the cylinder smoothly.

ii) Secondly the efficiency of an internal combustion engine depends upon the compression ratio defined as the ratio of the gaseous volume in the cylinder at the end of the sucktion stroke to the volume at the end of the compression stroke of the piston.

Higher the compression ratio better is the efficiency of the engine.

However if the compression ratio exceeds a certain limit the increased temperature of combustion causes spontaneous combustion of the fuel air mixture even before sparking. This is called 'pre-ignition'. Further it may also happened that the last unburnt portion of the fuel-air after sparking undergoes spontaneous 'self ignition' causing explosive violence. The explosion produces a shocking wave followed by a characteritic metallic sound called knocking. Knocking not only decreases the efficiency of the engine but also causes huge loss of energy and damage to piston and cylinder. The cause of knocking is characteristic of the fuel composition and engine design. It has been observed that

- The knocking tendency decreases with increase in compactness of the molecule, double bond and cyclic structures. Hence the trend of fuel constituents to known can be represented as
- straight chain paraffins > branched chain paraffins> olefins>cyclo paraffins> aromatic.
- Among the straight chain paraffins knocking increases with increase in chain length. Thus knocking property of straight chain paraffins can be graded as:

n-heptane > n- hexane > n-pentane > n-butane

Maximum power is derived from gasoline when it burns silently and relatively slowly. Under certain engine condition combustion first begins slowly and then whole of the unburn fuel may burn rapidly with formation of pressure wave leading to knocking.

Octane Rating – Octane rating or octane number is a scale of rating or classification of gasoline fuel depending on its knocking tendency. This process was first developed by Edger in 1926.

It was observed that n-heptane when used as a fuel in I.C (internal combustion) engine knocks very badly and hence it was assigned an antiknock value of zero. On the other hand iso-octane (2,2,4 trimethyl pentane) as a fuel has a great resistance to knocking and was rated 100 as its.

Anti knock value

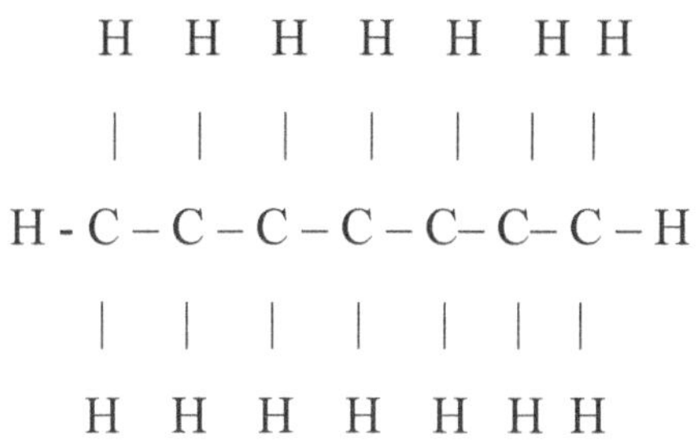

n– heptane

Anti knock value = 0

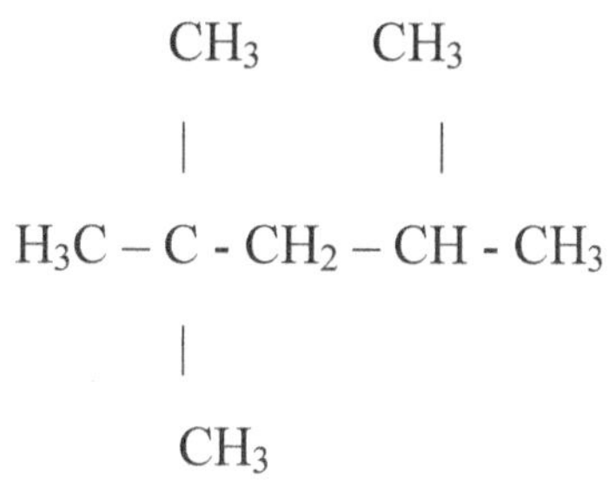

Iso-octane (2,2,4 trimethyl pentane)

Antiknock value = 100

Thus octane number or octane rating of gasoline or any other I.C engine fuel is the percentage by volume of iso-octane in a mixture of iso-octane and n-heptane, which matches the fuel under test in knocking tendency. For example a "80-octane" fuel will show the same combustion characteristics as that of a 80 : 20 mixture of iso-octane and n-heptane. Thus higher the octane number greater is the antiknock property and better is the fuel. Octane number of cracked gasoline is higher than straight run gasoline due to presence of higher percentage of unsaturated and cyclic hydrocarbons in the former.

Table : Octane number of common hydrocarbons

Hydrocarbon	Octane number	Hydrocarbon	Octane number
Benzene	106	N-pentane	62
Isopentane	90	N-hexane	26
Cyclohexane	77	Alcohol	95
2-methyl pentane	71		

The octane rating for motor gasoline in India is 83 and 93 where as in USA and Canada the two grades of motor gasoline have octane number 94 and 100 (premium). The aviation gasoline have octane number more than 100 which can also be achieved by adding an antiknock agent to gasoline of octane number 100.

Antiknocking agents:

Although the antiknock quality of a gasoline depends upon the type of hydrocarbons present but the octane rating of commercial gasoline can be improved by addition of certain chemical compounds. The most effective and common additive is tetra ethyl lead (TEL) which is used in a mixture of organic halides. The halides are added to prevent deposition of lead by forming volatile lead halides.

Paraffins and naphthenes have better lead susceptibility than aromatics. Other antiknock agents are tetramethyl lead (TML) and tertiary butyl acetate.

Unleaded Petrol

A unleaded petrol is one where enhancement of octane rating is made without addition of lead compounds. This can be done by

i) Reforming or structural rearrangement

ii) Addition of an antiknocking agents.

The octane number of gasoline is raised with high octane additives like isopentane, iso-octane ethyl benzene, isopropyl benzene, methyl tertiary butyl ether (MTBE). MTBE provides oxygen from its ether group which is used for combustion of petrol in IC engines and simultaneously reducing extent of peroxy compound formation.

Use of unleaded petrol permits the automobiles to work with catalytic converter attached to its exhaust. A catalytic converter contains rhodium as catalyst that checks pollution by converting toxic gases (CO and NO) to harmless gases (CO_2 and N_2) and oxidises unburnt hydrocarbon to CO_2 and H_2O.

(Lead poisons the rhodium catalyst by destroying the active sites).

Diesel Engine Fuels

In a diesel engine the fuel is ignited by application of heat and pressure. A diesel engine fuel should ignite below its compression temperature and these should be as short an induction lag as possible. A longer induction lag will cause accumulation of too much fuel in the cylinder, which will eventually burn with an undue rapidity. A rapid rise in cylinder pressure will cause a 'diesel knock'. The delay period is related to nature of hydrocarbon present in the fuel. The preferred hydrocarbons in order of increasing induction lag in knocking tendency.

n-paraffin < Olefins < naphthalene < iso-paraffin < aromatics

The above gradation indicates the ignition quality of a diesel fuel is highest with straight chain hydrocarbons.

Cetane Number:

The ignition quality of diesel engine fuel is measured in a standard engine by matching against blends of two referrence fuels and expressed as cetane number. A n-paraffin cetane (n-hexadecane $C_{16}H_{34}$) is given 100 cetane number while an aromatic compound namely d-methyl naphthane is rated as zero cetane number. If the given fuel matches with a 40/60 blend of cetane and n-methyl naphthaline it is rated as cetane number 40.

For high speed engines the diesel fuel required is of cetane number 45-60. While for low speed of the rating of fuel is of cetane number 25-30.

CH_3-$(CH_2)_{14}$-CH_3

CETANE

n-hexadecane(cetane no.100)

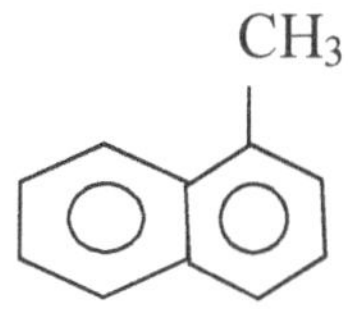

α-methyl naphthalene

CETANE no.0

The cetane number of diesel fuels rich in aromatics and isoparaffins can be enhanced with addition of antiknock compounds like alkyl nitrates (amyl and butyl nitrate), carbamates (isopropyl-n-methyl-n-nitrocarbamate), peroxides (ditertiary butyl peroxide) and metal-organic compounds). However these additives perform better in straight-run fractions to cracked fractions.)

Thus the characteristics of a hydrocarbons to be used as a good diesel fuel is just reverse for a good petrol fuel.

Tabel: Comparative study for petrol and diesel oil

Sl..	Gasoline petrol	Diesel oil
1.	It is a fraction obtained between 40-120 ^{0}C and a mixture of hydrocarbons C_5H_{12} to C_9H_{20}	It is a fraction obtained between 250-320 ^{0}C and is a mixture of hydrocarbons $C_{15}H_{32}$ to $C_{18}H_{38}$.
2.	Used in Internal combustion engines and required spark ignition.	Used in internal combustion engines and ignited by compression.
3.	Combustion requires a simple carburetor.	Combustion requires a more expensive fuel injection device.
4.	Ignition quality is measured by octane number	Ignition quality is measured by cetane number.
5.	Knocking is caused by pre-ignition	Knocking is caused by induction lag (delay ignition).
6.	Exhaust gases contain higher amount of pollutants	Exhaust gases contain lesser amount of pollutants
7.	Fuel consumption per unit of power production is more (Lower thermal efficiency).	Fuel consumption per unit of power production is less. (Higher thermal efficiency).
8.	TEL, Diethyl tellurides used as antiknocking agent.	Alkyl nitrites, acetone used as anti knocking agent.
9.	It is costlier.	It is cheaper.

Kerosene as a Fuel

Kerosene mainly used as a domestic purposes as stove oil and illuminant. It is obtained by fractional distillation of petroleum (crude oil) at temperature 180 to 250 ^{0}C having hydrocarbons from C_{10} to C_{16}. Elemental analysis shows its composition as C=84%, H=16% and S less than 0.1%. Its calorific value is about 11,100 Kcal/Kg with specific gravity of 0.75 to 0.85. It has a high boiling point and burns with a non-luminous flame when mixed with air. As a jet engine fuel kerosene is injected in to the compression chamber and mixed with excess of air before. It is ignited as a jet engine fuel the sulphur content of kerosene should be low which is done by action of ammonium polysulphide and conc. H_2SO_4 or washing with alkalies.

Table : Comparative study of liquid fuels

Sl. No.	Gasoline (petrol)	Diesel Oil	Kerosene Oil
1.	It is a mixture of hydrocarbons C_5 to C_8	It is mixture of hydrocarbons C_{15} to C_{18}.	It is a mixture of hydrocarbons C_{10} to C_{16}.
2.	Obtained from the fraction between 40-120 ^{0}C	Obtained from the fraction between 250-320 ^{0}C	Obtained from the fraction between 180-250 ^{0}C
3.	Used as a fuel in internal combustion petrol engines.	Used as a fuel in internal combustion diesel engines.	Used as a domestic stove illuminant, jet engine fuel and for making oil gas.
4.	Calorific value is about 11,250 Cal/gm	Calorific Value is about 11,000 Cal/gm.	It is calorific value is about 11,100 Cal/gm.

GASEOUS FUEL

Gaseous Fuels

Gaseous fuels are preferred over solid and liquid fuels due to their ease of storage, transport, handling, ignition and above all for its high calorific value. Gaseous fuels are free of mineral impurities and have consistency in quality. Hence gaseous fuels are widely used both for industrial and domestic purposes.

Classification: Gaseous fuels can be classified based on its mode of occurrence and method of production as follows:

Fuel gases occurring in nature:

- Natural gas and its other variations.
- Methane from coal mines.

Fuel gases made from solid fuels.

- Producer gas
- Water gas

- Coal gas
- Blast furnace gas
- Wood gas
- Gases from other gasification processes, including substitute natural gas (SNG).

Fuel gases made from petroleum

- Refinery gases
- Liquified petroleum gas (LPG)
- Gases from oil gasification process

Fuel gases made by fermentation of organic waste.

Gaseous fuels which have a calorific value above 4000 $KCal/Nm^3$ is called a rich gas. Hydrocarbon gas and coal gas are rich gas. On the otherhand gaseous fuels are known as lean gas when its calorific value is below 1500 $Kcal/Nm^3$. Blast furnace gas and producer gas are examples of lean gas.

Natural gas:

Natural gas is a mixture of paraffinic hydrocarbons, mostly occurs vicinity of coal mines and oil fields. The main component of natural gas is methane along with varying proportions of ethane, propane, butane and pentane. Small quantities of iso-paraffins and naphthenes may also be present. The vapors of the hydrocarbons excluding methane and ethane can be separated out as condensate. When natural gas contains very less (condensate$<$15 g/m^3) amount of these recoverable condensate it is called a dry gas. If the amount of recoverable condensate is appreciable (above 50 g/m^3) it is called as wet gas. Wet gas is commercially treated (compression, oil scrubbing and adsorption with activated charcoal) to recover the valuable condensate known as natural gasoline which is used as a blending agent for motor fuels. The residual gas after recovery of natural gasoline is called stripped natural gas, to distinguish it from dry gas. Similarly when natural gas comes out along with crude oil it is called associated gas (as in Bombay High)

Uses

1. Natural gas is widely used as a fuel for domestic and industrial heating as a pipeline gas.
2. Natural as is compressed and used as automobile fuel, a substitute for petrol or diesel.
3. Used as source for manufacture of carbon black(a filler for rubber) and hydrogen (for ammonia synthesis).
4. Used for synthesis of a number of chemicals.
5. Microbiological fermentation of methane to produce synthetic protein used as an animal feed.

Compressed Natural gas:

Natural gas is compressed to a pressure as high as 1000 atmosphere and hence known as compressed natural gas or CNG. It has a high calorific value of about 9000 to 11,200 Kcal/Nm^3. Natural gas is neither corrosive nor toxic. Its high ignition temperature and narrow flammability range makes it the safest fuel compared to other fuels. Due to its low energy density CNG is used as an automobile fuel a substitute for petrol and diesel. A steel-cylinder containing 15 kg of CNG contains about 20 m^3 of natural gas at 1 atm pressure is mounted in vehicles for the purpose. Natural gas has a high concentration of methane hence on combustion, CNG engines releases no sulphur or nitrogen gases. Emission of CO, CO_2 and hydrocarbons are also less by 70 to 90%,10% and 40-60% respectively than petrol or diesel engines for automobiles. Moreover, higher octane value of CNG reduces the knocking problem of a vehicle.

Advantage of CNG

i) The higher ignition temperature of CNG than gasoline and diesel make former (CNG) a much safer fuel.

ii) Operation cost of CNG is much lower than diesel or gasoline.

iii) Emission of CNG engine contains no unregulated pollutants (eg. Smoke, SO_X, NO_X, benzene etc).

iv) Being lighter than air, CNG always rises into the atmosphere without contaminating soil and water.

v) Considering the availability and infrastructure CNG qualifies to be one of the most prominent fuel both from life cycle emission and vehicle exhaust emission view point.

Liquified Petroleum Gas (LPG):

Gaseous fuels containing C_3 – (propane) and C_4(butane) hydrocarbons can be liquefied at room temperature by application of moderate pressure. The liquefied hydrocarbons can also be conveniently stored and transported in light pressure steel cylinders. Hence these are known as bottled gas or liquefied petroleum gas or LPG.

It is prepared from wet natural gas associated gas or refinery gas. The high calorific value, high specific gravity, high air requirement and low flame speed has made LPG superior to other gaseous fuels. A little amount of odorant like mercaptans or sulphides is added to odorless LPG to give warnings of gas leak.

Two grades of LPG are bottled for domestic and industrial use separately. LPG rich in C_3 hydrocarbons (90% propane) with least amount (<10%) of butane, isobutene ethane and ethylene (<2%) is used for domestic heating purpose, while LPG containing 80-90% butane-isobutane with propane-propylene(<20%) and ethane-ethylene(<2%) is used as industrial fuel. Special burners are available for use of LPG in domestic and industrial use.

Table:Calorific value of gaseous fuels

Sl. No	Type	Gross CV Kcal/Nm^{2}dns	Theoretical air required Nm^3
1	Pure butane	30,680	30.94
2	Pure propane	23,670	23.80

LPG is also used as a fuel in internal combustion engines of automobiles as a substitute for gasoline and diesel.

Advantages of LPG

i) The calorific value of LPG is approximately 3 times that of other gaseous fuels.

ii) Portability in steel cylinders, easy and safe transport makes its use possible in remote areas.

iii) Flexibility in control.

iv) Cleanliness in storage, handling and use.

v) As it is free from CO possess less health hazards.

COMBUSTION

Combustion is a chemical reaction which involves complete burning of a substance in excess of air or oxygen with liberation of heat (exothermic process) and light at a rapid rate.

For example:

$$C_{(s)} + O_2 \xrightarrow{combustion} CO_{2(g)} + \Delta H$$

Where $\Delta H = -97 Kcal$

For proper and complete combustion the substance (reactants) has to be brought to its ignition (kindling) temperature which may be defined as "**the minimum temperature required to ignite and burn the substance with further addition of heat from outside**".

Generally during combustion many solid fuels on ignition simply glow and get incandenscent while gaseous fuels on ignition always produce flames during combustion.

Combustion Calculation : Combustion calculation helps in determining the quantity of air/oxygen required for complete combustion of a unit quantity of fuel. The calculation is based on the following facts:

i) Determination of the stoichiometric (or theoretical) quantity of air or oxygen required to completely burn a known quantity of fuel and

ii) The actual amount of air /oxygen used for the purpose.

The elementary principles on which combustion calculation is based is explained as follows:

i) Substances always combine in definite proportions which are determined by the molecular masses of the reactants and products.

Explanation:

$$C_{(s)} + O_2 \rightarrow CO_{2(g)}$$

Mass proportion	12	32	44
Volume proportion	1	1	I

Thus by mass proportion C : O_2 : CO_2 is 12:32:44

Also $2H_{2(g)} + O_{2(g)} \rightarrow 2H_2O_{(g)}$

Mass proportion	2 x 2	32	2 x 18
Volume proportion	2	1	2

That is mass proportion is H_2 : O_2 : H_2O : : 4 : 32 : 36

Thus from balanced chemical equation mass of oxygen required for complete combustion can be easily calculated.

Following are the balanced equation for common components of any fuel.

$S + O_2 \rightarrow SO_2$

$CO + O_2 \rightarrow CO_2$

$CH_4 + 2O_2 \rightarrow CO_2 + 2H_2O$

$2C_2H_6 + 7O_2 \rightarrow 4CO_2 + 6H_2O$

$C_2H_4 + 3O_2 \rightarrow 2CO_2 + 2H_2O$

$2C_2H_2 + 5O_2 \rightarrow 4CO_2 + 2H_2O$

Nitrogen or ash present in the fuel being non-combustible matter do not consume any oxygen.

ii) According to Avogadro's Law 22.4 lts of any gas at STP (i.e. 0 ^{0}C and 760 mm pressure) occupies 1 mole of mass.

Thus at STP 22.4 lts of C will weight 12g (molar mass)

Or 22.4 lts of CO_2 will weigh 44 g (molar mass).

iii) Composition of air is as follows:

	% O_2	% N_2	% air /O_2
By weight	23	77	4.35
By values	21	79	4.76

iv) Molecular mass of air is {(32 x 0.23) + (2.8 x 0.777)} = 28.94 g mol^{-1}

v) Minimum O_2 required = Theoretical O_2 required – O_2 present in the fuel.

Minimum O_2 required should be calculated on the basis of complete combustion. In case of incomplete combustion, the product contains CO. In such cases the excess oxygen is calculated after subtracting the amount of oxygen required to burn CO to CO_2.

vi) The mass of dry flue gas formed should be calculated by balancing the C content of the fuel and that of the flue gas.

vii) The mass of any gas can be converted to its volume at certain temperature and pressure using gas equation PV= nRT where P, V and T are the pressure, volume and temperature of the gas in atmosphere, litres and Kelvin scale respectively. n represents number of mole of the gas and R being the gas constant.

viii) Oxygen present in the fuel is present in combination with hydrogen (as H_2O). This combined hydrogen is non-combustible hence does not take part in combustion. The rest of hydrogen called available hydrogen only takes part in the combustion reaction.

$$4H + O_2 \rightarrow 2H_2O + Heat$$

$$H_2 + \frac{1}{2}O_2 \rightarrow H_2O + Heat$$

Mass properties, 2 16

Since 1 part of hydrogen combines with 8 part of oxygen.

The available hydrogen = Mass of hydrogen - $\left(\frac{Mass\ of\ oxygen}{8}\right)$.

∴ Theoretical amount of oxygen required for

complete combustion of 1 kg fuel = $\left[\frac{32}{12}\times C + 8\left[H - \frac{0}{8}\right] + S\right]$ kg,

where C, H, S and O represent masses of carbon, hydrogen, sulphur and oxygen respectively per kg of fuel.

As the percentage of oxygen in air by mass is 23, so theoretical amount of oxygen required for complete combination of 1 kg of

fuel = $\frac{100}{23}\left[\frac{32}{12}\times C + 8\left(H - \frac{0}{8}\right) + S\right]$ kg.

Rclationship between Mass Analysis and volume analysis:

Volume proportions from masses is calculated as follows:

Let four gases A, B, C and D are taken in amount a,b,c and d. Their molecular masses are m_1, m_2, m_3 and m_4 respectively.

$\therefore$ Volume proportion of the gases are $\dfrac{a}{m_1} : \dfrac{b}{m_2} : \dfrac{c}{m_3} : \dfrac{d}{m_4}$

$$\therefore \text{ Total volume} = \frac{a}{m_1} + \frac{b}{m_2} + \frac{c}{m_3} + \frac{d}{m_4} = M \text{ (say)}$$

$$\text{Then \% of gas A} = \frac{a/m_1}{M} \times 100$$

$$\text{\% of gas B} = \frac{b/m_2}{M} \times 100$$

$$\text{\% of gas C} = \frac{c/m_3}{M} \times 100$$

$$\text{\% of gas D} = \frac{d/m_4}{M} \times 100$$

Similarly : From volume analysis, the mass of each gas can be determined by simply multiplying volumes of each gases with its molar masses and their proportional mass is determined and then from that the percentage by mass of each gas is determined.

Analysis of Flue gas :

The mixture of gases coming out of the combustion chamber of ovens and furnaces is known as flue gas. An analysis of the gas composition informs about the extent of combustion of the fuel. For efficient combustion the fuel is mixed with sufficient quantity of air/oxygen.

Hence the proportions of CO_2, CO and O_2 in the flue gas is measured.

- Presence of CO in the flue gas indicated about either incomplete combustion or inadequate supply of oxygen. Hence supply of oxygen is to be increased.
- Presence of excess of O_2 in the flue gas informs that though the combustion is complete the supply of O_2 is in excess. Hence supply of O_2 is to be shortened and the proportion of fuel to air or O_2 supply is to corrected.
- Presence of excess of both CO and O_2 in the flue gas indicates about irregular and non-uniform combustion of fuel. This may be due to the construction of the furnace. Some parts of it get excess air/O_2 supply and in other parts the supply is inadequate.

The analysis of the gas is carried out in Orsat's apparatus, where the volume of CO_2, O_2 and CO formed measured. Orsat's apparatus consists of a jacketed water measuring burette which is connected in series to a set of three absorption bulbs each through a stop-cock. The other end is connected to a three way stop-cock, which is further attached to a U-tube packed with fused calcium chloride. The lower end of the burette is attached to a water reservoir with the help of a long rubber tubing. The level of water in the water reservoir can be raised or lowered by raising or lowering the water reservoir.

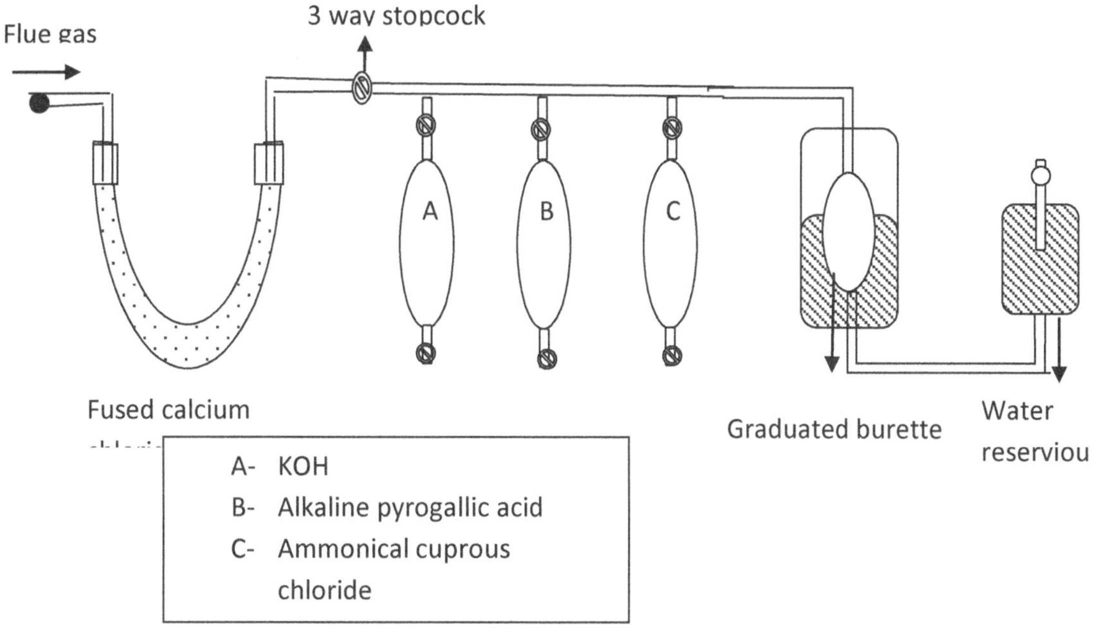

Fig: Orsat Apparatus

The absorption bulbs are filled with solutions like potassium hydroxide, alkaline pyrogallic acid and ammonical cuprous chloride solution for absorption of CO_2, O_2 and CO respectively from the flue gas.

Working :

1. The whole Orsat's apparatus is thoroughly cleaned, stoppers are greased and tested for air tightness. The three absorption bulbs A, B and C are filled with their respective solutions are closed with stop cocks. The water jacket and the leveling reservoir are filled with water.
2. The three way stop cock is opened to the atmosphere and the water reservoir is raised till the burette is completely filled with water and air is excluded from the burette.
3. The three way stop cock is now connected to the flue gas supply through a $CaCl_2$ U-tube and the reservoir is lowered to draw the gas in to the burette for analysis. However this sample gas being mixed with the air present in the apparatus, it is expelled to the atomosphere by opening the threeway stopcock followed by raising the reservoir. The process is repeated for three to four times to ensure complete removal of the air present in the apparatus.
4. Finally flue gas is sucked into the apparatus and its volume is adjusted to 100ml at the atmospheric pressure. This is done by adjusting the height of the reservoir till the level of water in the reservoir and burette becomes equal and is 100ml. The three-way stopcock is then closed.
5. The stopcock of absorption bulb A is opened and moisture free (absorbed in $CaCl_2$ U-tube) flue gas is then passed through the absorption bulb A containing caustic potash solution (250 gm of KOH in 500 ml of boiled distilled water) by raising the reservoir. The gas is again sent to the

burette. The process is repeated several times to ensure complete absorption of CO_2 by KOH solution by raising and lowering the reservoir.

The unabsorbed gases are finally taken back to the burette by raising the liquid in the bulb A till it is completely filled. The stopper of bulb A is closed. The level of unabsorbed gas is noted after equalizing the water level in the burette and water reservoir. The decrease in volume gives the volume of CO_2 in 100 ml of the flue gas sample.

6. Now the stopcock of bulb B containing alkaline pyrogallic acid (25g pyrogallic acid + 200 g of KOH in 500 ml of distilled water) is opened. The absorption process is repeated as in bulb A. The volume of gas absorbed given the volume of O_2 in 100 ml of flue gas sample.

   Similarly, the flue gas is passed through absorption bulb C containing ammonical cuprous chloride solution (100g cuprous chloride + 125ml liquor ammonia + 375 ml water) for analysis of CO present in the 100 ml flue gas sample.

The gas remaining in the burette after removal of CO_2, O_2 and CO is taken as nitrogen.

Producer Gas

The combustible gaseous mixture obtained by passing a mixture of air and steam over a bed of incandescent coal and coke is known as producer gas and the plant for preparation of the gas is called gas producer or simply producer. The main composition of this gas is carbon monoxide (25%), nitrogen (70%) and carbondioxide (4%) with small amount of H_2, CH_4 and O_2. However, the composition may vary slightly depending upon the carbonaceous material used as the gas producer can use low grade coal, wood waste bark etc as fuel. Producer gas is considered to be the cheapest industrial fuel. Due to presence of large amount of non-combustible nitrogen its calorific value is low (1300 Kcal/m^3).

The furnace consists of an air tight cylindrical furnace of 10-15 feet in height. Its inner side is lined with fire bricks with an outer steel casing. At the bottom of the furnace there is an air inlet and also an outlet for ash removal. At the top it has cup and cone feeder. The producer gas formed is removed through an upper side outlet.

Air or a mixture of air and steam is passed over coke and coal mixture at 1000^0C The reaction that take place in the producer leading to the manufacture of production gas can be explained with following zones.

a) **Ash zone :** It is the lowest zone of the furnace and is about 0.8m in thickness. This zone mainly consists of ash.Primarily the ash protects the grate (of the producer plant) from intense heat of combustion. Secondly the temperature of air-steam mixture is increased as they pass through the ash zone.

b) **Combustion zone :** This zone is present above the ash zone. Here oxidation of C takes place producing CO and CO_2. The reactions being exothermic temperature of this zone rises to 1100 ^{0}C. This zone is also called oxidation zone.

$$C + O_2 \rightarrow CO_2 + 97\ Kcal$$

$$C + \frac{1}{2}O_2 \rightarrow CO + 29.5\ Kcal$$

c) Reduction Zone :

The CO_2 produced under gas reduced with red hot coke in this zone producing free H_2 and CO.

$$CO_2 + C \rightarrow 2CO \quad --- \quad 36\ Kcal$$

$$C + H_2O \rightarrow CO + H_2 \quad --- \quad 29\ Kcal$$

$$C + 2H_2O \rightarrow CO_2 + 2H_2 \quad --- \quad 19\ Kcal$$

All these reactions being endothermic the temperature decreases to 1000^0C. If no steam is supplied the temperature of the producer rises high enough to melt the ash and refractory lining. To avoid this air mixed with a little steam is supplied.

d) **Distillation Zone:** This is the upper most zone where the downcoming coke/coal is heated by the outgoing gases (producer gas). The temperature of the zone is between 400-800 ^{0}C. The heat given by the outgoing producer gas and heat radiated from the reduction zone helps to distill the fuel, thereby volatile matter of coal is added to outgoing gases.

The produce gas is poisonous in nature. Although producer gas is combustible, it has low calorific value. It also heavier than air and insoluble in water.

Uses

The advantages of this lean fuel are

i) Low price and ease of production have made this gas an industrial fuel.

ii) Mainly used in metallurgical operations.

Water gas (or Blue Gas):

Water gas is also called Blue gas as it burns with a blue flame. Water gas or blue gas is a combustible gas of medium calorific value. It consists of nearly equal proportions of CO and H_2 along with minor amount of CO_2 and N_2 . The equipment used to produce the gas is called water gas generator.

The gas is produced by passing steam over red hot coke/coal at a temperature of about1000^0C.

$C + H_2O \rightarrow CO + H_2 - 28\ Kcal$ (Endothermic)

The above reaction is strongly endothermic. Thus, large quantity of heat is required to maintain the temperature. This is achieved by carrying out the production of water gas in a blow and run process. In the first half heat is generated and the temperature of the bed is raised by hot air blast followed by formation of water gas with absorption of heat.The gas generator consists of a cylindrical steel cylinder lined with refractory bricks. It has a cup and cone feeder at the top and a side opening for taking out the water gas produced. Near the bottom portion the furnace is provided with separate inlet pipes for passing air and steam and also with an outlet for the ash formed.

Air is passed over red hot coke or coal when carbon of coke burns to produce CO_2.

$C + O_2 \rightarrow CO_2 + 97\ Kcal$ (Exothermic)

Since this is an exothermic process, the temperature of the fuel bed increases to 1000 ^{0}C. After attainment of required temperature (1000 ^{0}C) the air blast is stopped, and steam is blown through the red hot coke to form CO and H_2.

$C + H_2O \rightleftharpoons CO + H_2 - 28.7\ Kcal$ - (1)

$C + 2H_2O \rightleftharpoons CO_2 + 2H_2 - 18.6\ Kcal$ - (2)

Formation of CO reaction predominates at high temperature while reaction (2) takes place at low temperature. The reactions being endothermic the temperature of the fuel bed decreases and formation of CO_2 predominates. Hence at this stage steam supply is cut off and air is again blown through the fuel bed to increase its temperature to 1000-1100 ^{0}C through the exothermic reaction and to facilitate co-formation ((reaction-1).

These two steps are repeated alternatively favorable condition to maintain proper temperature and for production of water gas.

Since the gas contains considerable quantity of CO it burns with a blue flame hence the name blue gas. Although it has very short flame but with high calorific value 2800-3100 Kcal/m^3.

Use:

i) Water gas is used as a fuel gas.

ii) As a source of hydrogen gas.

iii) As a raw material for Fischer-Tropsch. Process for manufacture of synthetic petrol.

iv) As an illuminating gas.

Carburetted Water Gas:

To increase the calorific value of water gas it is mixed with gaseous

hydrocarbons generated through cracking of crude oil. The carburetted water gas is enriched water gas with increased calorific value to about 4500 KCal/m^3. The composition of this gas is H_2 (35%), CO (25%) saturated and unsaturated hydrocarbons (35%) with minute quantities of N_2 and CO_2 (5%).

A water gas generator attached to a Carburettor super heater, hydraulic main and purifier respectively is used for production of Carburetted water gas. The water gas produced in the water gas generator is led through the carburetter. At the same time oil usually a light petroleum fraction is sprayed into the carburetor, which gets vaporized and cracked. In the carburetor water gas mixes with cracked oil vapors which is then passed through super heater where cracking process gets completed. The gas is then led through hydraulic main and a purifier. The product gas, is called carbureted water gas.

Uses:

The carbureted water gas possesses high calorific value and used mainly for illuminating and heating purposes.

Coal Gas (COKE OVEN GAS):

Coal gas is obtained by high temperature Carbonization of coal. Although it is a colourless gas but burns with a long smoky flame. It is lighter than air and its average composition is H_2 (40%), CH_4 (32%), CO(7%), with minute proportion of ethylene, acetylene N_2 and CO_2 having a calorific value of about 4900 $KCal/m^3$.

Coal gas is manufacture either in coke ovens in horizontal or vertical retorts by heating coal rich in volatile matter (30-40%) at a temperature of 1300 ^0C in absence of air.

$$Coal \xrightarrow{1300\ ^0C} Coke + Coal\ gas \uparrow$$

(In absence of air)

The gas obtained from gas making retorts is called coal gas which the gas obtained as a by product from coke ovens is named as coke oven gas.

The coal gas thus evolved is scrubbed by passing through a hydraulic main, where some tar and ammonia gets condensed. The hydraulic main also serve as a water seal against back passage of the gas to retort.

Remaining tar and ammonia present in the gas are separated by condensing the gas in a water cooled heat exchanger called condenser. The cooled gas is then passed through a washing tower under pressure and further chilled to separate out naphthalene. The chilled gas is then passed through another scrubber containing creosote oil which dissolves benzol, toluene and other aromatic compounds present in the gas.

The gas is then treated with moist Fe_2O_3 to remove any hydrogen sulphide present in it:

$Fe_2O_3 + 3\ H_2S \rightarrow Fe_2S_3 + 3H_20$

The Fe_2S_3 thus formed is subjected to aerial oxidation and converted to Fe_2O_3 to be reused.

$2Fe_2S_3 + 3\ O_2 \rightarrow 2Fe_2O_3 + 6S$

The purified coal gas (coke oven gas) is stored over water in gas holders.

Use:

- Coal gas is widely used as a town gas mainly for heating and illuminating purposes.
- It is also used for providing reducing atmosphere in metallurgical operations.
- It is a cheapest raw material for synthesis of ammonia.

OIL GAS:

Oil gas is obtained by cracking of Kerosene oil. Oil gas has a characteristic golden yellow color and burns with a smoky flame. Its average composition is CH_4 (25-30%), H_2(50-55%), CO(10-12%) and CO_2(3%) with a calorific value of 4500-5400 Kcal/m^3.

The plant consists of a strong cast iron retort, enclosed in a coal fired furnace. A bonnet is fitted at the mouth of the retort, through a molten lead seal. The other end of the bonnet is connected to a hydraulic main through a water seal. The prepared oil gas is stored in gas holder which is connected to the hydraulic main through a pipe with a testing tap facility.

The retort is heated and a thin stream of kerosene oil is continuously allowed to drip on the redhot bottom of the retort. The oil on coming in contact with red hot bottom, immediately gets cracked into a number of lower gaseous hydrocarbons.

$$C_{12}H_{26} \xrightarrow{Cracking} CH_4 + C_2H_6 + C_2H_4 + C_2H_2 + C_3H_8 + C_3H_6 + C((Tar)$$

Kerosene oil

The gaseous mixture thus formed passes out through the bonnet cover to hydraulic main, where tar gets condensed and removed.

The quality of the gas formed is estimated by withdrawing some gas through the testing tap. A golden yellow colour of the gas depicts a good quality oil gas which ensures completion of cracking process. However required colour of the gas can be obtained by adjusting the supply of air and kerosene oil. The oil gas thus formed is stored over water in gas holders. The yield of oil gas in this process is $\approx 40-50\%$.

Uses:

Oil gas is mainly uses as a laboratory gas and use to increase the calorific value of water gas as carbureted water gas.

Table :Comparative study of gaseous fuels

Sl. No	Fuel	Average composition	Manufacturing	Calorific valuein (Kcal/m^3)
1	Natural Gas	CH_4=70-90% C_2H_6=5-10% H = 3% CO + CO_2=Rest		12,000- 14,000
2.	Producer Gas	CO = 22 – 30% H_2 = 8-12% N_2 = 52 – 55% CO_2 = 3%	**Oxidation** $C + O_2 \rightarrow CO_2$ $C + \frac{1}{2}O_2 \rightarrow CO$	1300

			Reduction $CO_2 + C \rightleftharpoons 2CO$ $C + HO_2 \rightleftharpoons CO + H_2$ $C + H_2O \rightleftharpoons CO_2 + H_2$	
3.	Water Gas	$H_2 = 51\%$ $CO = 4\%$ $N_2 = 4\%$ $CO_2 = 4\%$		2800
4.	Carbureted water Gas	$H_2 = 34\text{-}38\%$ $CO = 23\text{-}28\%$ $N_2 = 2.5 – 5\%$ $CO_2 = 0.2 – 2.5\%$ Hydrocarbon = 30 – 48%	$C + H_2O \rightarrow CO + H_2$ $Oil \xrightarrow{Cracking} Hydrocarbon$	3500 - 4450
5.	Coal Gas	$H_2 = 40\%$, $CH_4 = 32\%$ $CO = 7\%$, $C_2H_2 = 2\%$ $C_2H_4 = 3\%$, $N_2 = 4\%$ $CO_2 = 1\%$ rest = 4%	$Coal \xrightarrow{1300\ ^0C} Hydrocarbon$ Coke + Coal gas	4900
6.	Oil Gas	$CH_4 = 25\text{-}30\%$, $H_2 = 50\text{-}55\%$ $CO = 11\%$, $CO_2 = 3\%$	$C_{12}H_{26} \xrightarrow{Cracking}$ $CH_4 + C_2H_6 + C_2H_4 + C_2H_2 + C_3H_8$	6700

Biogas (Gobar gas):

Biogas is produced by bacterial degradation of organic waste in absence of free oxygen (anaerobic degradation). Natural gas also a biogas which is produced by a long period of decay of vegetable and animal matter. Biogas is obtained from carbonaceous agricultural waste of plant or animal origin like animal dung , straw, hay, fruit peelings, municipality sewage etc.

The anaerobic fermentation of biomass takes place in two stages. First the complex organic compounds are decomposed into simple organic acid and secondly these acids are later converted into methane and carbon dioxide. The acid forming and methane forming bacteria must remain in a state of dynamic equilibrium.

The production of biogas (gobar gas) is made in a plant called biogas plant (gobar gas plant) which consists of a well-shaped underground tank called digester covered with a dome shaped roof both made of bricks and cement. The dome of the digester is also used as a gas holder and is attached to outlet pipe and a gas valve. The digester is connected to one chamber each at both the sides. The left chamber used as an inlet chamber while the chamber to the right hand side is used as an outlet chamber where spent dung slurry gets collected.

Slurry made by mixing cattle dung and other waste material with water is fed into the digester through the inlet upto the cylindrical top level. In about 50 to 60 days time the dung starts undergoing fermentation by bacteria through anaerobic degradation and the biogas thus produced gets collected under the dome. Slowly the produced biogas exert pressure on the slurry in the digester and pushes the spent lye to the overflow tank via outlet chamber. The spent lye is used as a very good manure.

The biogas thus produced is mainly used for domestic heating and illuminating purposes. Its calorific value is about 5300 Kcal/m^3. It is interesting to note that heat generated by direct burning of 1kg of cattle dung is 23.4 Kcal while the same amount of cattle dung when converted to gobar gas can supply 188 Kcal of heat.

Composition: The average composition of biogas is

methane (55%), H_2(7.4%), CO_2(35%) and N_2(2.6%).

Advantages:

Presence of high percentage of methane makes biogas an excellent domestic fuel. Neither it possesses any poisonous gas nor any storage is required for it. Biogas can directly be supplied through pipeline to a maximum distance of 10 meters from the plant. It burns without smoke, causing no pollution. Thus it is a boon to village ladies who usually suffer from eye sight problem and lung's disease arising from the pollution caused by smoky fuels used to cook food. Biogas production is very economical too.

Non-Petroleum Fuels:

Benzol : It is obtained as a by-product of coal carbonization having a distillation range of 70-150 ^{0}C. It is also obtained from fractional distillation of light oil. Its average composition is benzene (70%), toluene (18%), xylene(6%) and other hydrocarbons. It has a very high calorific value of about 10,200 Kcal/kg. Benzol has a very high antiknock quality with octane rating 87-90. Due to high initial boiling point and freezing point at 55 ^{0}C, benzol can not be used as a motor fuel along (never in cold countries). So benzol is used as a blending agent for gasoline. By blending 50%motor benzol, octane number of gasoline may be raised from 70 to 80.

Power Alcohol:

Ethyl alcohol is used as a liquid fuel for internal combustion engines and is referred to as Power Alcohol. It is never used as a prime fuel rather blended with gasoline in the ratio 1:4. Ethyl alcohol has a high octane rating 90. So when blended it increases the octane number (antiknock property) of gasoline. The calorific value of ethyl alcohol is low (7000 Kcal/kg). Moreover due to presence of low boiling fraction in ethyl alcohol starting problem is cold engines are found when ethyl alcohol alone is used as a fuel. So ethyl alcohol always used as a blend with gasoline.

Manufacture:

In India alcohol is manufactured from molasse's, the dark brown mother liquor left after crystallization of sugar from concentrated sugar cane juice. It contains about 50-60% sucrose and fructose. This mother liquor(molasses) is diluted with water to bring down the sugar concentration to 10 to 12%. To it some nutrients (ammonium sulphate and ammonium phosphate) are added. Followed by dilute sulphuric acid to maintain the pH between 4 to 5. Then adequate quantity of yeast (an enzymes mixture) is added. Temperature of the reaction mixture is maintained at 30 ^{0}C. The invertase (of the yeast) converts sucrose to glucose and fructose.

$$C_{12}H_{22}O_{11} + H_2O \xrightarrow{Invertase} C_6H_{12}O_6 + C_6H_{12}O_6$$

Sucrose Glucose fructose

The zymase (of the yeast) converts glucose and fructose into ethyl alcohol and carbon dioxide as follows.

$$C_6H_{12}O_6 \xrightarrow{Zymasee} 2C_2H_5OH + 2CO_{2(g)}$$

The froth produced during the fermentation process is due to liberation of CO_2. The total fermentation process takes 36 to 38 hours to complete. The fermented liquid known as wash contains about 18-20% alcohol. It is subjected to fractional distillation to get rectified spirit which contains 90-95% of alcohol. The rectified spirit is there after digested with lime for 48 hours followed by distillation to get absolute alcohol which contains 100%. The middle fraction of the distillate is collected.

Advantages:

1. Blending of alcohol to petrol increases the octane number. Hence alcohol blended petrol possesses better antiknock properties.
2. Alcohol-petrol blend have lesser starting problem.
3. Alcohol petrol blend have a tendency to absorb any traces of moisture present.

Disadvantages:

1. Alcohol on mixing with petrol lower the calorific value of the blend as the calorific value of alcohol is two-third that of petrol.
2. Alcohol is easily oxidized to acid and hence may cause corrosion.

3. The ratio of flow of blend is to be increased to 1.56 times that of petrol by altering the carburetor jet size.
4. Since air required for complete combustion of blend is less than the air required for petrol alone, the air entering the cylinder is to regulated by modifying air inlet or by operating choke.

QUESTIONS –

1.Define fuel and its classification.

2.What do you mean by calorific value of fuels?

3.What are the characteristics of a good fuel?

4.Distinguish between gross calorie value and net calorific value of a fuel.

5.Distinguish between solid and gaseous fuel.

6.Write different varieties of coal formed inside earth.

7.Write the name of the highest ranking of coal.

8.The jelly like substance formed during the first stage of coalification of wood is named as what?

9.What is ignition temperature of fuel ?

10.What is the main purpose of proximate analysis of coal

11.What are the characteristics of a good quality coal ?

12.What is carbonization of coal?

13.What is the difference between caking coal and cokingcoal ?

14.Why coke is preferred to coal in metallurgical processes?

15.Why uniform coal size is favored as a fuel?

16.What do you understand by fixed carbon. How it affects the calorific value of coal ?

17.What is the effect of temperature on carbonization of coal ?

18.What is regenerating principle used in Otto-Hoffman's byproduct oven?

19.What are the byproducts recovered from Otto-Hoffman's byproduct oven?

20.What is petroleum. Write a short note.

21.Classification of liquid fuel. Explain

22.Give a comparisons between paraffin base and asphalt base crude.

23.What is straight run gasoline?

24.What are the products obtained from fractional distillation of crude oil?

25.What do you understand by cracking?

26. What is thermal cracking?

27. What is catalytic cracking. Explain its merits.

28. What is synthetic petrol?

29. Name the process used to convert coal to petrol.

30. What is the catalyst used in Fischer-Tropsch?

31. Write a comparative study of straight run, cracked and polymer gasoline.

32. What is knocking?

33. What do you understand by anti knocking?

34. What is octane rating from petrol?

35. What is citen number?

36. Why petrol cannot be used in diesel engine?

37. Why tetra ethylead is added to gasoline for internal combustion engine.?

38. Why water gas is superior to producer gas?

39. What is unleaded petrol?

40. Write a short note on Bomb calorimeter.

41. Explain drift theory of formation of coal.

42. What do you know by cleaning of coal.

43. Write short note on-

    a)Sour gasoline

    b)SweetPetrol

44. Write a note on analysis of coal.

45. Write a note on Bee-hive coke oven?

46. Write a note on Origin of petroleum.

47. Explain CNG and its merits.

48. What is coal gas?

49. What is gobar gas. Explain its merits.

50. Write the characteristics of good fuel.

Photochemistry

Chapter at a glance

- *Laws of Photochemistry*
- *Quantum yield*
- *Primary and Secondary photochemical processes*
- *Kinetics of photochemical reactions*
- *Photosensitized reaction*
- *Fluorescence and Phosphorescence*
- *Masers and Lasers*
- *Photosynthesis*
- *Oxygen liberation and Carbon dioxide consumption in photosynthesis*
- Various theories *of mechanism* of photosynthesis.

Photochemistry is the branch of chemistry that deals with photochemical reaction and other chemical effect of light. The reactions which occurs as a result of energy acquired by interaction of electromagnetic radiations with matter are known as photochemical reactions. For the study of photochemistry the term light includes ultraviolet, visible and infrared regions of spectrum i.e; wavelengths ranging from 1000 to 10,000A°.Incase of photochemical reactions when light interact with matter, some energy of incident radiation is converted intofree chemical energy of the products and when the light source is removed the system returns to its ground state. Thus photochemical reactions are usually accompanied by an increase in free energy. Ozonization of oxygen, decomposition of ammonia or polymerization of anthracene are a few examples of photochemical reactions.

On the other side thermal or dark reactions are caused by energizing molecular collisions bought about by heat. These collisions are usually accompanied by a decrease in free energy.

Although light consists of electromagnetic waves but for the purpose of photochemistry it is best to regard light as composed of photons or light quantum.

The energy contained in each quantum of light is

$$E = h\vartheta$$

Where h= Planck's constant = 6.667×10^{-7}erg Sec

ϑ= Vibration frequency of light

E = Energy

It is the most fundamental equation of photochemistry which clarifies that higher the frequency greater is the energy per quantum

$$E \alpha \vartheta$$

So a quantum of violet light has much more energy than a quantum of red light. Thus the former is more photochemically active. Similarly ultraviolet light is more photochemically active then visible light. So sometimes violet and UV portion of sunlight or any source of light is called Actinic light.So for photochemical reaction to occur in laboratory a regular source of UV or visible light like arc light, mercury vapour light and Sunlight are used.

Absorption of light

When light falls on any surface, the incident ray suffer various changes like reflections, absorption and the transmission

Let, Ir = The intensity of incident radiation reflected

Ia = The intensity of incident radiation absorbed

I_t = The intensity of incident radiation transmitted

Io =The total intensity of incident radiation

According to the principle of conservation of energy

$$Io = Ir + Ia + I_t$$

Out of these Ia the intensity of radiation that absorbed, is related to photochemistry. The absorbance of light i.e;electromagnetic radiation is governed by the following two laws

i) Lambert's Law

ii) Beer's Law

Lambert's Law – The law states that when a beam of monochromatic radiations passes through a homogeneous absorbing medium, equal fraction of incident radiations are absorbed by successive layers of equal thickness of the light absorbing substance.

∴Rate of decrease in intensity of incident radiation with thickness of the medium = $-\frac{dI}{dx}$

Where I = intensity of radiation after passing through a layer of the thicknessdx

If k is the proportionality constant

$$-\frac{dI}{dx} \propto kI$$

or, $\frac{dI}{I} = -k.dx$

on integration $\int \frac{dI}{I} = -k \int dx$

or, $\log_e I = -kx + c$

where c= integration constant

when x = 0 and I = Io (Intensity of incident radiation)

$\log_e I = kx + \log_e I_o$

or, $\log \frac{I}{Io} = -kx$

or, $I = Io.\ e^{-kx}$

This is the mathematical relationship of Lambert's law.

Beer's Law (1852) : Beer extended the Lambert's law to solutions. This law when applied to solution is known as Beer- Lambert's Law. According to this law when an absorbing substance is in the gaseous state or dissolved in a transparent solvent, the absorption co-efficient is proportional to the concentration.

Mathematically

$$-\frac{di}{dc} = k'I$$

Where c = Molar concentration of the solution

k'= Absorption coefficient of the solution.

On integration and proceeding similarly

$$I = Io.e^{-k'c}$$

Changing from natural to Briggsian Logarithm

We have

$$I = Io \times 10^{-0.4343k'c}$$

Or $\frac{I}{Io} = 10^{-\alpha' c}$ Beer's Law

α' = Molar extenction coefficient = a constant

i.e; $\alpha' = 0.4343k'$

From Lambert's Law

$\frac{I}{Io} = 10^{-\alpha x}$ Lambert's law

Combining these two equation

$\frac{I}{Io} = 10^{-acx}$ Beer-Lambert's law

c is usually expressed in gm. mole/lt.

And x in cm. The value of 'a' depends on units of concentration.

When $\frac{I}{Io} = \frac{1}{10}$and c = 1

Thus Beer-lamberts law become

$$\frac{1}{10} = 10^{-acx}$$

or, acx = 1

∵(c= 1)

$a = \frac{1}{x}$

$a \propto \frac{1}{x}$

Thus 'a' is equal to the reciprocal of thickness in cm of one molar solution through which the intensity of light is reduced to $\frac{1}{10}$ of its value. 'a' is known as molar extinction co-efficient.

Limitations of Beer-Lambert's Law :

1. Beer-Lambert's Law is applicable for dilute solutions only, deviations set in with solutions of high concentration.
2. Beer-Lambert's law is not obeyed if radiations used in not monochromatic.
3. Change of temperature also affects the absorption. Increase of temperature shifts the absorption towards longer wavelength (bathochromic effect)

Effects of absorptions of light

When a system absorbs light radiation several different phenomenon are observed.

i. If there is no possible electronic, atomic or molecular change which can utilize the exact amount of energy associated with the quantum of radiation, there will be no interaction, neither physical nor chemical change will occur.

ii. If there is some change within the molecule corresponding to that of quantum, there may be a transfer of energy and the quantum is absorbed. As result of absorption the primary effect in all photo process is

   - To increase the thermal energy or
   - Electrons are excited and cause activation i,e; electronic excitation.

iii. In photo processes when the radiant energy is very high it causes electronic excitation to ionize atoms and to produce photoelectric effect.

iv. When the light energy is not sufficient to eject electrons to ionize atoms , it promotes electrons to higher energy level, But at higher energy level (excited state) the electrons are unstable and return to the normal state by re-emitting a part or total absorbed energy in front of light radiation. If the electrons return to the normal state instantaneously the effect is termed as

Fluoroescence.If the emission of radiation takes place after a time lag, the effect is known as Phosphorescence(delayed fluorescence).

v. If the electronically excited system collides with another atom or molecule, then the energy will be transferred from the excited system to the atom or molecule. This types of transition is responsible for photochemical reactions.

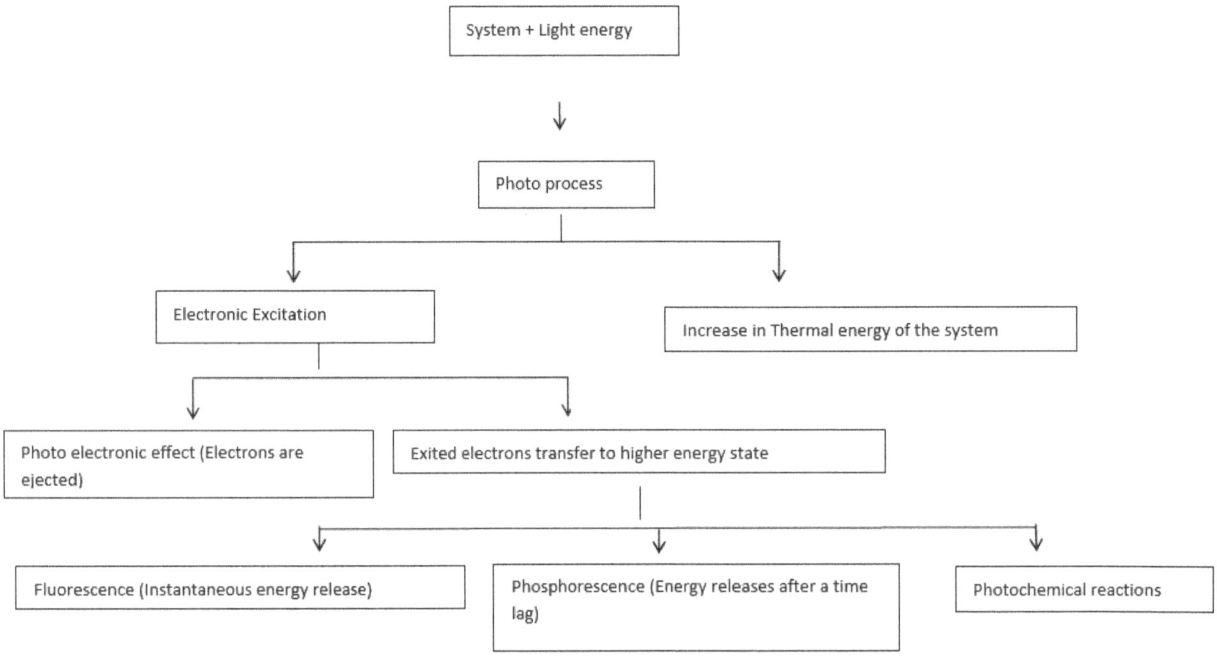

Fig:Flowchart depicting effects of light absorption by a system.

Laws of Photochemistry

There are three fundamental laws of photochemistry which govern the actions of light on systems

1. **Grothus- Draper Law –** This law is also referred as the first lawof photochemistry. The law states that only radiations which are absorbed by the reacting system are effective in producing a photochemical change.

   Although photochemical reactions occur from absorption of radiation but the total light absorbed is not responsible for the photochemical change. Only light covering a particular small range of wavelength is responsible for the photochemical change. For example when white light is passed through an alkaline solution of cupric tartarate, the dissolved substance undergoes photochemical reduction to cuprous oxide, Absorption spectrum measurements indicate infrared, orange and ultraviolet light also absorbed by the solution, but only UV light , covering a small range of wavelengths brings about the Photochemical reduction. Thus it can be said that UV light is the effective light for the photochemical change.

   Limitations

   a) The Grothus- Draper Law is based on qualitative facts only.

2.Second law of Photochemistry or Law of photochemical equivalence or Stark – Einstein Law-

The law sates that each molecule taking part in a chemical reaction induced by exposure to light absorbs one quantum of radiation causing the reaction (effective light).

The law implies a 1:1 relationship between the number of reacting molecules and the number of quanta of light absorbed applicable to primary process only.

Thus according to Stark-Einstein Law

Energy absorbed per mole is called Einstein of radiation and is equal too

$$E = Nh\vartheta$$

$$= Nh\frac{c}{\lambda}$$

Where c = velocity of light = 3.0×10^{8}m/s

λ = Wavelength of absorbed radiation

h = Planck's constant = 6.625×10^{-34}Js

N = Avogadro's number = 6.023×10^{23}/mol

$$\text{Thus } E = \frac{6.023\times 10^{23}\times 6.625\times 10^{-27}\times 3\times 10^{8}}{\lambda}$$

$$= \frac{0.1196}{\lambda}\text{J/mol}$$

$$E = \frac{0.1196}{4.184\times 10^{10}\,\lambda}\text{ k cal /mol}$$

$$= \frac{2.859\times 10^{5}}{\lambda}\text{ k cal/mol ....(A)}$$

λ is expressed in Angstrom units

According to the law of photochemical equivalence one mole of absorbing substance should decompose for every $(2.859 \times 10^{5})/\lambda$ kcal/ mole of radiation absorbed and the photochemical equivalent of 1kcal is thus $\lambda/(2.859 \times 10^{5})$ molecules.

Thus from eqn(A)

$$E \,\alpha\, \frac{1}{\lambda}$$

$$\text{Energy absorbed } \alpha \frac{1}{wavelenth}$$

So the energy absorbed per mole is greater for shorter wavelength of light.

That is for violet light λ = 4000 A° , E = 71 kcal

Red light λ = 7500A°, E = 38 kcal

So violet portion of the light is more photochemically active than of with longer wavelength.

3) <u>Third law of Photochemistry or Platnikow Law</u>

The law is formulated byPlatnikow states that , "if a system can undergo a change in dark and in the presence of light, then the total reaction observed will be the sum of effects produced due to darkness as well as light.

∴Total chemical change = change in dark + change in light

Although no quantitative proof of the law has been produced till date.

<u>Photochemical Processes</u>

According to Bodenstein all photo chemical reactions take place in two distinct processes.

- Primary process - It is the process when the reacting molecules/ atoms absorbs light radiation, got activated and forms excited molecule/atom.

$X_{(molecule/atom)} + h\vartheta_{\text{(photon of energy)}} \rightarrow X'_{\text{(Excited molecule or atom)}}$

$Cl_2 + h\vartheta \longrightarrow 2Cl^*$

All primary process are photochemical in nature.

- Secondary process – The excited (activated) molecule or atom or free radical formed in the primary process undergo photochemical change to yield the product.

$H_2 + Cl^{\circ} \rightarrow HCl + H$

Secondary process can occur in absence of light radiation. Hence secondary processes are not photochemical processes.

<u>Classification of photochemical process</u>

Photochemical process can be classified as given below.

i. Dissociation or pholytic reaction :some example of as follows:

$4\,NH_{3(g)} \xrightarrow{h\vartheta} N_{2(G)} + NH_2.NH_{2(g)} + 4H_{2(g)}$

$2\,HI_{(g)} \xrightarrow{h\vartheta} H_{2(g)} + I_{2(g)}$

$2\,HBr_{(g)} \xrightarrow{h\vartheta} H_{2(g)} + Br_{2(g)}$

ii. Double decomposition :

Some examples of this type of photochemical reactions are :

$C_6\,H_{14} + Br_2 \xrightarrow{h\vartheta} C_6H_{13}Br + HBr$

$CH_3ClCOOH + H_2O \xrightarrow{h\vartheta} CH_2OHCOOH + HCl$

iii. Addition reaction : For example Bromination of tetraphenylethylene as follows :

$Br_2 + (C_6H_5)_2C{=}C(C_6H_5)_2 \xrightarrow{h\vartheta} (C_6H_5)_2CBr.CBr(C_6H_5)_2$

iv. Polymerization reaction – Dimerization of anthracene occurs as follows :

$$2\,(C_{14}H_{10}) \xrightarrow{h\vartheta} (C_{28}H_{20})$$

v. Chain reaction –

$$H_2 + Cl_2 \xrightarrow{h\vartheta} 2HCl$$

vi. Rearrangement reaction :

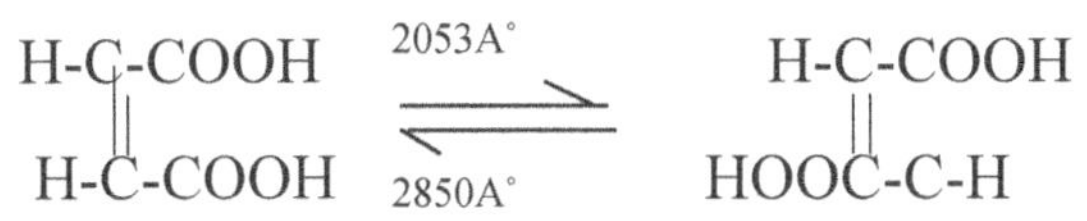

Maleic acid　　　　　　　　　　Fumaric acid

vii. Photosensitized reaction : For example

$$H_2C_2O_4 \xrightarrow[UO^{+2}]{h\vartheta} H_2O + CO_2 + CO$$

Quantum Yield or Quantum efficiency

Quantum yield or quantum efficiency is defined as the number of moles of the substance undergoing photochemical change per Einstein of radiation absorbed. It is donated as φ

$$\varphi = \frac{No.of\ molecules\ reacting\ in\ a\ given\ time.}{No.of\ quanta\ of\ light\ absorb\ in\ the\ same\ time.}$$

$$= \frac{No.of\ moles\ reacting\ in\ a\ given\ time.}{No.of\ einstein\ absorbed\ in\ the\ same\ time.}$$

$$= \frac{Rate\ of\ chemical\ reaction}{no.of\ einstein\ absorbed}$$

Thus the result of photochemical change are conveniently expressed in terms of quantum yield φ. According to stark-Einstein law photochemical equivalence of the value of φ should always be one as in the primary process with absorption of one quantum only one molecule undergoes dissociation and the products does not enter in further reaction, there will be 1:1 ratio between the number of quantum absorbed and number of reacting molecules.

However in the secondary process changes the quantum yield value from one. It is because in some processes number of molecules undergoing chemical change due to only one photochemically activated molecule is no more one but more or less than one, thereby changing the value of φ

Thus for complicating circumstances and secondary processes the value of φ in not equal to one.So the law of photochemical equivalence holds good for primary process strictly.

Information about mechanism of photochemical reaction is obtained from the value of quantum efficiency φ. Determination of the value of quantum yield can be obtained from the following.

i) Number of moles of reacting substance at a given time – This is done by standard analytical methods used for chemical kinetics.

ii) Number of Einsteins of light absorbed: This is determined with the help of an instrument known as actinomer, hence the process is known as actinometry. It gives data about the number of quanta falling per unit time on the reacting vessel (incident light) and the number of quanta emitting per unit time from the reaction vessel (emitted light). From this quantum yield φ is calculated. Depending upon the value of φ the photochemical reaction are classified as below.

Table – Quantum yield and classificalion of photochemical reactions

Example	REACTION	Quantum yield
Reactions where quantum yield is one $\varphi = 1$		
	$Br_2 + C_6H_{12} \rightarrow C_6H_{11}Br + HBr$	1
	$H_2 + O_2 \rightarrow H_2O_2$	1
	$Cl_2 + O_2 \rightarrow Cl_2O_2$	1
	$SO_2 + Cl_2 \rightarrow SO_2\,Cl_2$	1
	$2NO_2 \rightarrow 2NO + O_2$	1
Reactions in which $\varphi < 1$		
	$2NH_3 \rightarrow N_2 + 3H_2$	0.2
	$H_2 + Br_2 \rightarrow (5100A^{\circ})\ 2HBr$	0.01
	$CH_3COCH_3 \rightarrow C_2H_6 + CO + CH_4$	0.2
	$CH_3CH \rightarrow CH_4 + CO$	0.5
	$2C_{14}H_{10} \rightarrow C_{28}H_{28}$	0.25
Reaction in which $\varphi > 1$		
	$H_2 + Br_2 - (2070 - 2500A^{\circ}) \rightarrow 2HBr$	2.0
	$H_2O_2 \rightarrow H_2 + O_2$	2.1
	$\frac{3}{2}O_2 \rightarrow 2O_3$	2.0
	$3O_2 \rightarrow 2O_3$	3.0

Reaction in which $\varphi \gg 1$	
$Cl_2 + H_2 \rightarrow 2HCl$	10^{4-5}
$Cl_2 + O_2 \rightarrow COCl_2$	10^3

- Reactions with quantum yield less than one or lower quantum yield.
- Reactions with quantum yield more than one or higher quantum yield.
- Reactions with very high quantum yield.

Reasons for high quantum yield :

i. In some reactions the primary step of absorption of radiation produces atoms or free radicals which initiate a series of chain reactions. The actual quantum yield in such cases depends on how far the chain reaction can proceed before termination.

ii. Formation of an intermediate product which is capable to perform as a catalyst and giving rise to high quantum yield.

iii. The reactions may be exothermic and the heat evolved many activated other molecules; thereby these molecules react without absorbing radiation.

iv. The active molecules produced by primary absorption may collide with other molecules and activating them, which in turn activate other reacting molecules.

Example

The reaction between hydrogen and chorine gases on exposer to light radiation of 4800A° gave very high quantum yield of the order $10^4\ to\ 10^6$.

$$H_{2(g)} + Cl_{2(g)} \rightarrow 2HCl_{(g)}$$

According to the mechanism of given by Nenrst (1918) the quantum of energy absorbed in the primary process is much more than required for primary excitation process. Hence the molecules carry excess energy and are capable of starting secondary process, resulting with higher quantum yield

$$Cl_2 + h\vartheta \rightarrow 2Cl^\circ \text{ (Primary Process)}$$

$$\begin{cases} Cl^\circ + H_2 \rightarrow HCl + H^\circ \\ H^\circ + Cl_2 \rightarrow HCl + Cl^\circ \end{cases} \text{(Secondary Process)}$$

The regeneration of Cl° after production of two HCl molecules permits propagation of the secondary reactions.The secondary reaction are capable of producing even in dark as a result there is possibility of formation of an infinite number of HCl molecules for absorption of a single quantum of energy in the primary process yielding a very high quantum yield.

Reasons for Low Quantum Yield :

i. Occurrence of some processes involving deactivation of excited molecules before they are able to react and form products.

ii. All absorbing molecules may not received sufficient energy to react.

iii. Energised molecules may lose their energy by collision with non energised molecules.

iv. Primary photochemical reaction may get reversed.

Example:

Photochemical combination between hydrogen and bromine to form hydrogen bromide has a quantum yield of about 0.1.

Primary process

$Br_2 + h\vartheta \rightarrow \quad 2Br\bullet$ - - - - - (1)

Secondary Process

$Br\bullet + H_2 \rightarrow HBr + H\bullet$ - - - - (2)

$H\bullet + Br_2 \rightarrow HBr + Br\bullet$ - - - - (3)

$H\bullet + HBr \rightarrow H_2 + Br\bullet$ - - - - -(4)

$Br\bullet + Br\bullet \rightarrow Br_2$ - - - —(5)

The following mechanism has been proposed for its low quantum yield.

Above reaction (1) and(2) being highly endothermic, hence takes place very slowly at ordinary temperature. Due to slow speed of reaction (2) most of the bromine free radicals recombine to give bromine molecule. As a result reaction (2) ,(3) and(4) cannot occur and the quantum yield is very low.

Considering the photochemical polymerisation of anthracene

$$2\,C_{14}H_{10} \underset{\text{Dark}}{\overset{\text{Light}}{\rightleftharpoons}} C_{28}H_{20}$$

the quantum yield is 0.6.The low quantum yield is due to occurrence of back thermal reaction.

Kinetics of Photochemical Reaction :

The kinetics of photochemical reaction have been studied by derivation of rate law using steady state approximation for transient species. Kinetics of some typical photochemical reactions are described here with.

1. **Photochemical synthesis of Hydrogen chloride.**

As discussed earlier this reaction is accompanied with exceptionally high quantum yield ranging from 10^4 - 10^6 A. According to the mechanism suggested by Nernst (1918) after photodissociation chlorine molecules the secondary process follows a chain reaction mechanism.

Primary process

$Cl_2 + h\vartheta \xrightarrow{K_1} 2Cl$ - - - (1)

Secondary process

Propagation:

$$Cl + H_2 \xrightarrow{K_2} HCl + H \text{ - - - -(2)}$$

$$H + Cl_2 \xrightarrow{K_3} HCl + Cl \text{ - - - -(3)}$$

Termination: $Cl + Cl \xrightarrow{(wall)\ K_4} Cl_2$ - - - - -(4)

Reaction (2) and (3) i.e; propagation step are responsible for conversion of a large quantity of H_2 and Cl_2 to HCl.

The rate of formation of HCl is

$\frac{d[HCl]}{dt} = K_2 [Cl][H_2] + K_3[H] [Cl_2]$ - - - - - (5)

But concentration of [H] and [Cl] are required for determination of rate from equation (1).Thus applying steady state approximation i.e; the concentration of free radical or intermediate is constant at any instant or ,

Rate of formation of reactive intermediate = Rate of its decomposition.

Rate of formation of[Cl] is

$\frac{d[Cl]}{dt} = K_1 I_{abs} + K_3 [H][Cl_2] - \{K_2[Cl\}[H_2] + K_4[Cl]\}$ - - - - (6)

The chain termination step i.e; reaction(4) may be a recombination of chlorine atoms in gas phase with the help of a third body(M or wall of the vessel)to carry away excess energy.

Apply steady state assumption

$d[Cl]/dt = 0$

or, $K_1 I_{abs} + K_3 [H][Cl_2] = K_2[Cl\}[H_2] + K_4[Cl]$ - - - - - (7)

Similarly for [H] we can give $\frac{d[H]}{dt} = 0$

or, $\frac{d[H]}{dt} = K_2[Cl][H_2] - K_3 [H][Cl_2] = 0$

or,$K_2[Cl][H_2]$ = K_3 $[H][Cl_2]$ - - - - (8)

∴By addition of eqn (7) and (8) we have-

K_1I_{abs} -$K_4[Cl]$ =0

$$[Cl] = \frac{K_1 I_{abs}}{K_4} \text{ ----(9)}$$

Substituting the value of [Cl] in eqn (8)

$$K_2\frac{K_1}{K_4}I_{abs}\,[H_2] = K_3[H][Cl_2]$$

$$\text{or, } [H] = \frac{K1\,.K2Iabs[H2]}{K3\,.K4[Cl2]} \text{ - - - - - (10)}$$

Now after substituting the value of [H] and [Cl] in we have

$$\frac{d[HCl]}{dt} = \frac{K_1.K_2.I_{abs}[H_2]}{K_4} + \frac{K1.K2Iabs[H2]}{K4}$$

$$= 2\,\frac{(K1\,.K2)\ Iabs\ [H2]}{K4}$$

Rate = $K'\ I_{abs}[H_2]$

$$\text{where } K' = \left(2\,\frac{K_1.K_2}{K_4}\right)\text{......(11)}$$

Thus Rate ∝ I_{abs}

Eqn (11) proves that rate is directly proportional to the intensity of absorbed radiation explaining the exceptional high quantum yield of this photochemical radiation.

2.<u>Photochemical Synthesis of Hydrogen Bromide</u> :

$H_{2(g)} + Br_{2(g)} \xrightarrow{h\vartheta} 2HBr_{(g)}$

Hydrogen bromide is synthesised photochemical when $H_{2(g)}$ and $Br_{2(g)}$ are exposed to a continuous spectrum of wavelength 5100A°.In the primary process of Br_2 molecule dissociative into Br atoms.

Primary process:

$Br_2 + h\vartheta \xrightarrow{K1} 2Br$-----------------(1)

Seccondary Process

propagation:

$Br + H_2 \xrightarrow{K2} HBr + H$ ......(2)

$H + Br_2 \xrightarrow{K3} HBr + Br$ .....(3)

$H + HBr \xrightarrow{K4} H_2 + Br$ ....(4)

Termination :

$Br + Br \xrightarrow{K5} Br_2$ ......(5)

HBr is formed by reaction (2) and (3) but removed by reaction(4)

∴The rate of formation of HBr

$\frac{d[HBr]}{dt} = K_2[Br][H_2] + K_3[H][Br_2] - K_4[H][HBr]$ ......(6)

For determination of value of [Br] and [H] steady state approximation of intermediate is applied

$\frac{d[H]}{dt} = 0 = K_2[Br][H_2] - K_3[H][Br] - K_4[H][HBr]$ ......(7)

and $\frac{d[Br]}{dt} = 0 = K_1I_{abs} - K_2 [Br][H_2 + K_3[H][Br_2]+ K_4[H][HBr] - K_5[Br]^2$ ......(8)

From eqn (7) and (8)

$[Br]= (\frac{K_1}{K_5} Iabs)^{1/2}$ ......(9)

Similarly from eqn(7) on rearrangement we have

$[H] = \frac{K_2[Br][H_2]}{K_3[Br]+K_4[HBr]}$ ---------------(10)

On substituting the value of[H] and [Br] in eqn(1)

the rate of formation of [HBr] is rate $= \frac{d[HBr]}{dt}$

$$=K_2[H_2]\sqrt{\frac{K_1}{K_5}Iabs} + \{K_3[Br_2]-K_4[HBr]\}\left(\frac{K_2[H_2]\sqrt{\frac{K_1}{K_5}I_{abs}}}{K_3[Br]+K_4[HBr]}\right)$$

$$= K_2[H_2]\sqrt{\frac{K_1}{K_5}Iabs}\left(\frac{K_3[Br]+K_4[HBr]+K_3[Br_2]-K_4[HBr]}{K_3[Br_2]+K_4[HBr]}\right)$$

$$= K_2[H_2]\sqrt{\frac{K_1}{K_5}Iabs}\left(\frac{2K_3[Br_2]}{K_3[Br_2]+K_4[HBr]}\right)$$

$$= K_2[H_2]\sqrt{\frac{K_1}{K_5}Iabs}\left(\frac{2}{1+\frac{K_4[HBr]}{K_3[Br_2]}}\right)$$

$$=2K_2[H_2]\sqrt{\frac{K_1}{K_5}Iabs}\left(\frac{[H_2]}{1+\frac{K_4[HBr]}{K_3[Br_2]}}\right) \quad ------------(12)$$

Thus Rate $\propto\sqrt{Iabs}$

Although the combination of H_2 with Cl_2 and H_2 with Br_2 consist of chain reaction, the quantum yield of the two photochemical reactions are widely different. It is to be noted that quantum yield of a photochemical reaction depends upon propagation step.

In case of $H_2 + Cl_2 \longrightarrow 2HCl$ $\varphi \simeq 10^4 - 10^6$.

The reaction is exothermic, hence formation of HCl takes place spontaneously leading to high quantum yield.

$H_2 + Br_2 \longrightarrow 2HBr \qquad \varphi = 0.01$

On the otherhand the propagation step of the above reaction, HBr is endothermic and takes place extremely slowly at ordinary temperature.

$Cl + H_2 \xrightarrow{K_2} HCl + H$ (exothermic)

$Br + H_2 \xrightarrow{K_2} HBr + H$ (endothermic)

As quantum yield depends upon the number of times the propagation steps are repeated before termination the quantum yield for HBr is less.

3.<u>Photochemical dissociation of HI :</u>

The dissociation of HI takes place in the region of 2000 - 2300A°. The primary photochemical step is.

$HI + h\vartheta \xrightarrow[K2]{} H + I$

Secondary process

$H + HI \xrightarrow{K3} H_2 + I$

$I + I \xrightarrow{K4} I_2$

Thus the rate of consumption of HI is

$-\frac{d[HI]}{dt} = K_1 . I_{abs} + K_2[H][HI] \quad \ldots\ldots\ldots(1)$

The rate of formation of[H] is

$\frac{d[H]}{dt} = K_1 I_{abs} - K_2[H][HI]$

Apply steady state approximation of[H] we have

$\frac{d[H]}{dt} = K_1 I_{abs} - K_2[H][HI] = 0$

$\therefore K_2[H][HI] = K_1 I_{abs}$

Substituting the value of $K_2[H][HI]$ in eqn(1), we have

$$-\frac{d[HI]}{dt} = 2K_1 I_{abs}$$

This shows that the absorption one photon leads to decomposition of two molecules of HI.Thus the quantum yield is 2.

$$\varphi = 2\frac{K_1 K_2}{K_1 Ia} = 2$$

However as reaction progress. Concentration I_2 increases the following thermal reaction takes place. As a result the quantum yield decreases.

$$H_2 + I_2 \xrightarrow{K4} HI + I$$

Hence the steady state approximation for[H] now becomes.

$$\frac{d[H]}{dt} = K_1 I_{abs} - K_2[H][HI] - K_4[H][I_2] = 0$$

$$[H] = \frac{K_1 Iabs}{K_2[HI] + K_4[I_2]}$$

Substituting the value of [H] in eqn(1) we have

$$-\frac{d[H]}{dt} = K_1 Iabs + K_2[HI]\frac{K_1 Iabs}{K_2[HI] + K_4[I_2]}$$

$$= K_1 Iabs\left(1 + \frac{1}{1 + \left(\frac{K_4[I_2]}{K_2[HI]}\right)}\right)$$

$$= -\frac{d[HI]/dt}{Iabs} = K_1 + \frac{K_1}{1 + \left(\frac{K_4[I_2}{K_2[HI]}\right)}$$

As the reaction proceeds$[I_2]$ increases but[HI] decreases, hence quantum yield decreases from the original value of 2.

Photochemical dissociation of HBr is similar to that of HI.

Photochemical Equilibrium :

In some photochemical reaction an equilibrium is established between reactants and products of decomposition.It is different from thermal reaction equilibrium that the forward reaction is photochemical while the reverse reaction takes place in absence of light.

$$\text{Reaction} \underset{light}{\overset{dark}{\rightleftarrows}} \text{Products}$$

The quantum yield of two processes in some cases are different.This photochemical equilibrium is known as photo stationary state. Some example of this type of photo stationary state is presented below.

i) Decomposition of Nitrogen dioxide:

Vapours of nitrogen dioxide when exposed to light of wavelength less than 3700A° decomposes to nitric oxide and oxygen, but the reverse reaction proceeds in dark.

$$2NO \underset{light}{\overset{dark}{\rightleftarrows}} 2NO + O_2$$

2) Di mineralization of anthracene :

Anthracene when dissolved in a solvent like benzene dimerises to dianthracene in presence UV radiation but the reverse reaction takes place in dark.

$$2C_{14}H_{10} \underset{light}{\overset{dark}{\rightleftarrows}} C_{28}H_{20}$$

Anthracene *Dianthracene*

3) Conversion of sulphur dioxide to sulphur trioxide

This reaction was studied by Cohen and Becker(1970)

$$2SO_2 + O_2 \rightleftarrows 2SO_3$$

In this case both forward and reverse changes takes place in UV light. The photo stationary state is independent of the temperature ranging from 500 - 800°C

Equilibrium Constant:

Consider following reaction which is light sensitive in forward direction while the reverse reaction takes place in dark.

$$A \underset{K_1 light}{\overset{K_2 dark}{\rightleftarrows}} B$$

Rate of forward reaction =K_1 I abs

Rate of reverse reaction = $K_2[B]$

At photostationary state

K_1 I abs =K_2 [B]

$$K = \frac{K_2}{K_1} = \frac{Iabs}{[B]}$$

Thus K = Photochemical equilibrium constant in proportional to the intensity of light absorbed.

Difference between Thermal and Photochemical Reaction –

Reaction can be classified as

1. Thermal or Dark reaction
2. Photochemical reaction

1) **Thermal reaction** – Thermal reaction are caused by energizing molecule collisions by heat and in absence of light, while the reactions which occurs as a result of energy acquired by interaction of electromagnetic radiations with matter are known as photochemical reaction .The following table lists some important important differences between these types of reaction.

Sl. No.	Thermal Reaction	Photochemical Reaction
1	Thermal reaction do not require light to proceed, it can take place even in dark.	Photochemical reaction can occur only in presence of light radiation.
2	These reactions involve heat change between reactants and products.the reactions may be endothermic or exothermic.	Photochemical reaction involve absorption of light radiation only.
3	All thermal reaction proceed with decrease of free energy(ΔG =-ve)	Photochemical reactions usually proceed through increase of free energy.

Molecular Excitation -

It is a known fact that most of the molecules contain even number of electrons in the ground state these electrons are paired.On excitation one of these electrons are paired.On excitation one of these electrons per molecule goes to higher energy level leaving the original energy level half filled.The spin orientation of these two electrons may be same or different.

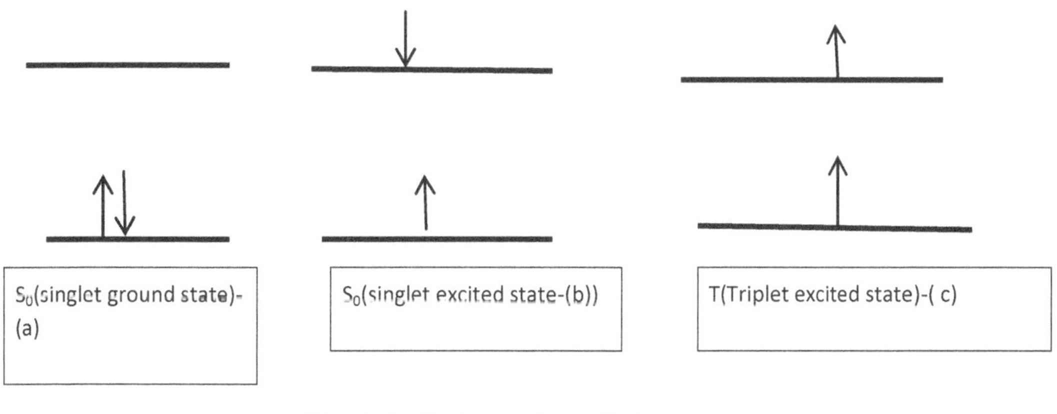

Fig:Spin Orientaation of electrons.

The spin multiplication of a state is governed by (2S+1) rule where S represents total electronic spin.

For Example:

In the ground state when the spins are paired like fig(a) the spin that is clockwise orientation of one electron is cancelled by the clockwise orientation of the other.

Thus $S = S_1 + S_2$

$\therefore$ Spin multiplicity = 1/2 -1/2 = 0

= 2S +1 = 1

i.e; spin multiplication is 1

or, The molecule is in singlet ground state(s)

2) On absorption of a photon of energy $h\vartheta$, one of the paired electrons goes to higher energy level. Let in the excited state the spin orientation of the electrons are parallel like fig(b).

a) Parallel(↑↑)

Thus spin multiplicity $2S+1 = S_1 + S_2 + 1$

$= 2(1/2 + 1/2) + 1$

$= 3$

Thus the molecule is triplet(T) in excited state.

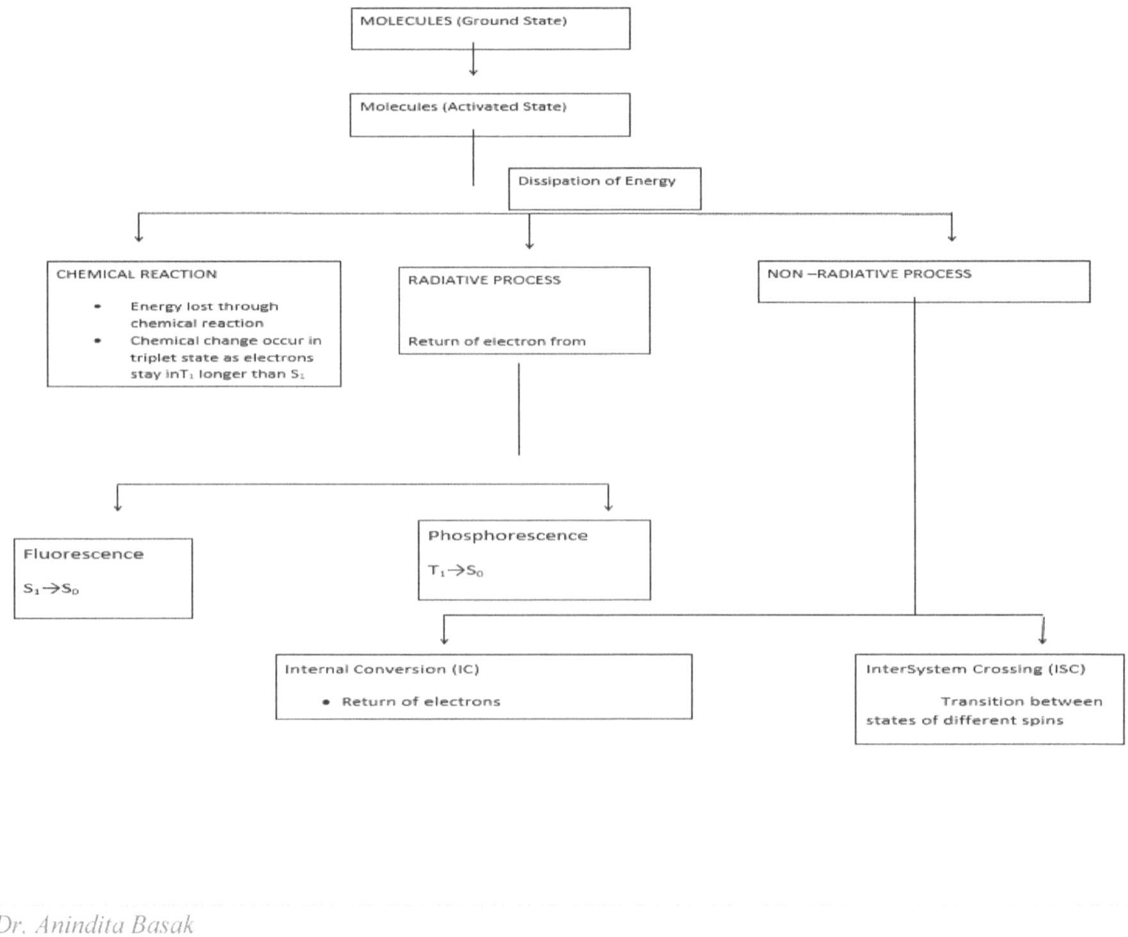

b) Antiparallel (↑↓)

In this case $S = 1/2 - 1/2 = 0$

$\therefore 2S + 1 = 2 \times 0 + 1 = 1$

Thus the molecule is in singlet(s) excited state.

Depending upon the quanta of photon($h\vartheta$) absorbed electrons can jump to any higher state giving us a series of singlet and triplet states.

Therefore we can express

S_o- Singlet ground state

$S_1, S_2, S_3, \ldots\ldots$ etc = first, second, third ,...... etc singlet excited state respectively.

Similarly, $T_1, T_2, T_3, \ldots\ldots$ etc are first, second ,third,...... etc triplet excited state respectively.

According to Quantum mechanism

- Singlet excited state is of higher energy than corresponding triplet excited state.

On in other words $E_{(S1)} > E_{(T1)}$

$$E_{(S2)} > E_{(T2)} \ldots\ldots$$

-Direct excitation of singlet ground state to first triplet excited state is forbidden it has to be routed through singlet excited state i.e; $S_0 \rightarrow S_1$ and $S_1 \rightarrow T_2$.

<u>Consequences of Light Absorption :</u>

When a molecule absorbs light energy ,its electron may get excited from singlet ground state (S_o) to singlet excited state Sn(where n =1,2,3,...) depending upon the energy of the absorbed light photon.In principle for every singlet excited state there exist a corresponding triplet excited state.As already discussed direct excitation of singlet ground state.So to triplet excited state is forbidden, it has to routed through corresponding single excited state.

The molecule is said to be activated when it is in excited state singlet or triplet.

Thus , $M_o + h\vartheta \rightarrow M^*$

where M_o = molecule in ground state.

M^* = molecule in the excited state.

Being unstable in the excited state the molecule returns to the ground state through following different processes

1. **<u>Non radiation transition -</u>**

This process includes transition from higher excited states likeS_3, S_2 orT_3 ,T_2 to the first excited state S_1 or T_1.As this transition does not involve emission of any type of radiation ,it is known as radiation less or non radiativetransition.The energy of the activated molecule is released in following ways.

i) *Internal Conversion*(IC) - In internal conversion the energy of the activated molecule is dissipated as heat through molecular collisions.The process occurs in less than 10^{-11} sec.

ii)*Intersystem Crossing* (ISC) - In this method the activated molecule loses its energy by transition between states of different spins i.e;

$S_2 \rightarrow T_2$ or

$S_1 \rightarrow T_1$ respectively as triplet excited state has lower energy hence longer life time (10^{-5} to 10^{-3} sec) than corresponding singlet excited state.

2.Radiative Transition -

The activated molecules return from singlet or triplet excited state to singlet ground state (S_o) by emission of radiations hence known as radiative transition. Depending upon the type of transition they are known as fluorescence and phosphorescence.

Singlet Excited State (S_1) $\xrightarrow{\text{Fluorescene}}$ Singlet Ground State (S_0)

And Triplet Excited State (T_1) $\xrightarrow{\text{Phosphorescence}}$ Singlet Ground State (S_0)

FLUORESCENCES

When a beam of light radiations of high frequency incident on certain substances, they emit light of different frequencies. The phenomenon is called Fluorescence and the substance is called fluorescent substance. The fluorescent substance stops emitting light as soon as the incident light is cut off.

Explanation-

When light is absorbed by a fluorescent substance it gets excited to higher electronic state.

$X + h\vartheta \rightarrow X^*$

Being unstable in the excited state the atom emits light radiations in several transitions return to lower states.

$X^* \rightarrow X^{1*} + H\vartheta'$ (secondary)

$X^{1*} \rightarrow X + h\vartheta''$

Where X^{1*} is the intermediate energy state between X and X^*.

It is clear that the emitted radiations are of different frequencies ($h\vartheta'$ and $h\vartheta''$) than that of incident radiation .The time taken for emitting the excess radiation is very short i.e; 10^{-9} to 10^{-4} sec, Thus fluorescence is regarded as a secondary process results from the primary process of absorption of a quantum of light ($h\vartheta$) by an atom or molecule(X).

Example of Fluorescent substance - Calcium fluoride, Uranium, petroleum, organic dyes like eosin and fluorescein, chlorophyll, quinine sulphate solution, vapours of sodium, mercury, iodine etcshows fluorescent with the phenomenon of light of different wavelengths for example.

Chlorophyll — with red light

Uranium glass — with green light

Fluorspar — with blue light and so.

Application-

- Fluorescent dyes mixed with coloured paints produce glow
- Aqueous fluorescent solution having intense green fluorescence use as a marker for leak detection or for see rescue operation or in industry.
- Used in fluorescent tubes for lightening purpose.

PHOSPHORESCENE

Phosphorescence is also a radiative process through which an activated molecules loses its energy and return to ground state. When light radiation of high frequency is incident on certain substances, they continue to emit light for sometime even after the incident light is cut off. This delayed fluorescence is called Phosphorescence .Depending upon the material, temperature and other factors the afterglow for fractions of second to hours to some days. The substances which show the phenomenon of phosphorescence is called phosphorescent substances.

Explanation-

Phosphorescence involves excitation of the electron to singlet excited state (S_1) followed by intersystem crossing to triplet excited state(T_1).The transition of electron from triplet excited state (T_1) to singlet ground state (S_0) by emission of energy is termed as phosphorescence. The transition from the triplet state is forbidden transition and is long lived as triplet state is of lower energy than singlet state.

$M + h\vartheta \rightarrow M^* \xrightarrow{slow} h\vartheta'$

Example-

Zinc sulphide, alkaline earth sulphide like CaS,BaS and SrS are phosphorescent substances.

Application -

Phosphorescent substances like ZnS is widely used as luminous clock dial and other substances which are made to glow in dark.

TABLE: Comparative study of Fluorescence and Phosphorescence.

Sl.No	FLUORESCENCE	PHOSPHORESCENCE
1	Fluorescence is the radiative emission of absorbed radiation in which the molecule in the excited singlet state i.e; $S_1 \rightarrow S_0$	Phosphorescence is also a radiative emission of absorbed radiation in which the molecule in excited triplet state returns to ground singlet state i.e; $T_1 \rightarrow S_0$
2	The transition $S_1 \rightarrow S_0$ is permitted and occurs in about 10^{-6}secs	The transition $T_1 \rightarrow S_0$ is forbidden transition and occurs in about 10^{-3} to 30 secs.

3	Fluorescence is exhibited by gases, liquids and solids.	Mostly Phosphorescence is exhibited in solids.
4	Fluorescence stops as soon as the incident light is cut off.	Phosphorescence continues even after the incident light is cut off. Hence phosphorescence is also known as delayed fluorescence.
5	The extent of fluorescence depends upon the nature of the solvent and presence of certain anions in solution	The time of emitted radiation depends upon nature of the substance, temperature and other variants.

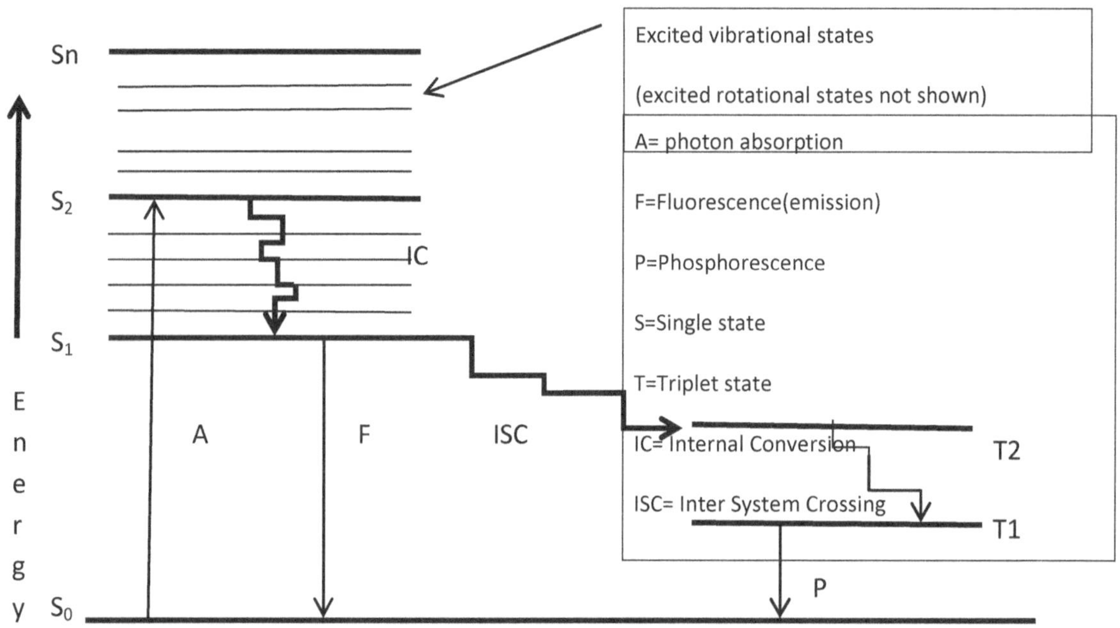

Fig: Jablonski diagram.

CHEMILUMINESCENCE

It is a phenomenon when visible light is given out by a chemical reaction taking place at a temperature at which black body will not give out visible radiation.

The glow fire flies gives out intermittent light due to chemiluminescence of protein luciferin.The protein luciferin present in body of glow flies oxidises in presence of an enzyme luciferase.

When phosphorous vapours are oxidised, it produces luminosity and the glow increases with decrease of pressure when more phosphorous vaporises.Thus oxidation of P_2O_5 and P_2O_3 produces chemiluminescence. Chemiluminescence is produces due to oxidation of pyrogallol by H_2O_2.

THERMOLUMINESCENCE

It is a phenomenon when a preraditated crystalline material is heated ,it emits light radiation of different wavelength. For example X-rays when absorbed by different crystals gives different coloured lights - Sodium chloride becomes yellow but Potassium chloride becomes blue.

This is due to absorption of light by electrons that have been released by X-rays and are trapped in the negative ion 'vacancies' in the crystal lattice. On heating the pre radiated crystals ,the trapped electrons are released and return to lower energy level by emitting light.

PHOTOSENTIZATION

The process of photosensitization was discovered by Nobel Laureate scientist Franck and Carioin.1922. It has been found that certain reactions which are not originally sensitive to light can be made photosensitive by addition of some foreign substances. The added foreign substances can absorb radiation when exposed to and initiates a photochemical reaction without itself, undergoing a chemical reaction .These substances are known as sensitisers and the phenomenon as photosensitization. Photosensitization is different from catalysis because photosensitsed reactions are spontaneous and involving an increase in free energy of the system. Some examples of photosensitizers include vapour of Hg ,Cd or use of uranyl sulphate, AgBr and so on.

Mechanism -

The role of photosentisers and reactants in a photosensitization reaction is similar to that of Donor(D) - Acceptor(A)system. The photosensitizer plays the role of a donor and absorbs the incident photons.

i) After absorption of photon the donor D' riches the singlet excited state which in turns gives triplet excited state D^3 via intersystem crossing(ISC)

ii) The donor atom D^3 at triplet excited state then collides with acceptor A(the reactant) and produces its triplet excited state A^3 the product and itself returns to ground state D^0.

iii) This happens if the triplet excited state of D i.e;D^3 is higher in energy than that triplet excited state of the reactant A i.e; A^3.

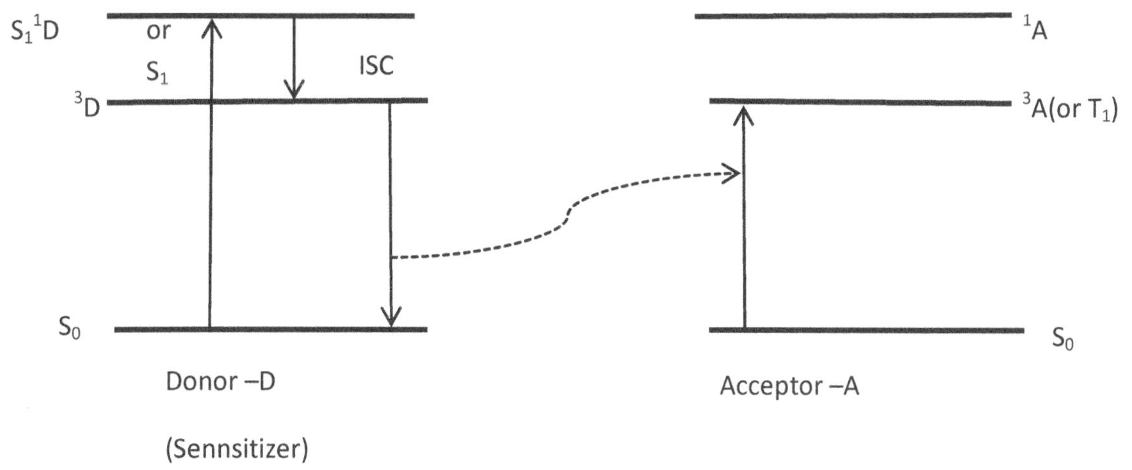

Fig: Mechanism of Photosensitization.

Thus the mechanism of photosensitization can be represented as follows:

$D + h\vartheta \longrightarrow D^1$

$D^1 \xrightarrow{ISC} D^3$

$D^3 + A \longrightarrow D + A^3$

iv) Instead of A^3 if D^3 gives the product the process is called quenching and A is known as quencher.

$A^3 \rightarrow$ Product (Photosensitization)

$D^3 \rightarrow$ Product (Quenching)

Examples:-

I - **Mercury as photosensitizer.**

Hg vapour absorbs photon when exposed to light of wavelength 2537A° and gets excited and initiates dissociation of hydrogen molecule by transfer of energy.

i) Photosensitized dissociation of H_2

$Hg + h\vartheta \rightarrow Hg^*$

$Hg^* + H_2 \rightarrow 2H + Hg$

ii) Photosensitized chain reaction

$Hg + h\vartheta \rightarrow Hg^*$

$Hg^* + H_2 \rightarrow 2H + Hg$

$H + O_2 \rightarrow HO_2$

$HO_2 + HO_2 \rightarrow O_2 + H_2O_2$

$HO_2 + H_2 \rightarrow H_2O_2 + H_2$

iii) Photosensitized reaction of H_2 with CO

$Hg + h\vartheta \rightarrow Hg^*$

$Hg^* + H_2 \rightarrow 2H + Hg$

$H + CO + X \rightarrow HCO + X$

$2HCO \rightarrow HCHO + CO$

$HCO + H_2 \rightarrow HCHO + H^*$

$2HCO \rightarrow (CHO)_2$

iv)Photochemical decomposition of NH_3 or PH_3

$NH_3 + h\vartheta \xrightarrow{2537Å} NH_2 + H$

$NH_2 + NH_2 \rightarrow N_2 + 2\,H_2$

$H+H \longrightarrow H_2$

v)Decomposition of Ozone

As explained above the mechanism can be sketched as follows:

$Cl_2 + h\vartheta \rightarrow Cl_2^* \rightarrow 2Cl$

$Cl + O_3 \rightarrow ClO_3^*$

$2ClO^*_3 \rightarrow 2Cl + 3O_2$

and partly polymerises

$2ClO_3^* \rightarrow Cl_2O_6$

ClO_3• radical will react with ozone

$ClO_3 + O_3 \rightarrow ClO_2 + 2O_2$

$ClO_2 + O_3 \rightarrow ClO_3 + O_2$

Silver halides (AgBr) in photographic plate are not affected by red or yellow light but in presence of cyanine dyes used as sensitisers can photosensitize the photographic plate.

II - **Cd as Photosensitizer**

Cd vapour also act as a photosensitizer at 3262A° wavelength and can photosensitize decomposition of hydrocarbons like methane, ethane and so on.

III-Uranylsulphate as photosentizer

Uranylsulphate on exposure to light radiation of 2540 to 3660A° can photo initiate the decomposition of oxalic acid.

$UO_2^{2+} + h\vartheta \rightarrow UO_2^*$

$UO_2^* + (COOH)_2 \rightarrow CO + CO_2 + H_2O + UO_2$

IV- Chlorophyll as photosensitizer

In photosynthesis process chlorophyll from the plant behaves as a photosensitiser to absorb sunlight between 6000 - 7000A° for food preparation.

$CO_2 + H_2O^{h\vartheta} \longrightarrow HCHO + O_2$

$6HCHO \rightarrow C_6H_{12}O_{6(\text{carbohydrate(food)})}$

PHOTOSYNTHESIS

Photosynthesis, the nature only photochemical process where species possessing green pigments chlorophyll, convert solar energy to chemical energy liberation of oxygen. Almost all living being depend directly or indirectly on the process of photosynthesis for food and breathing.

In photosynthesis all plants, algae, some bacteria (blue green cyan bacteria) and some protists contain the green pigment 'chlorophyll' produce carbohydrate (food) and oxygen using carbon dioxide ,water and sunlight as follows.

$6CO_2 + 12\ H_2O + \text{Sunlight} \rightarrow C_6H_{12}O_{6(\text{carbohydrate})} + 6O_2 + 6\ H_2O$

- During photosynthesis carbon dioxide is procured through 'stomata' (the tiny pores present on plant leaves) and liberates oxygen is also released to the atmosphere through these stomata.
- Water is obtained by plants through roots and delivered to the leaves by vascular plant tissue following the process of osmosis.
- Sunlight is absorbed by Chlorophyll

The several structures present inside chloroplast and their function is outlined below

Outer and Inner membrane - These are protective covers around chloroplast.

Stroma - It is the area between the grana and filled with dense fluid present inside chloroplast, where conversion of CO_2 to sugar takes place.

Thylakoid - It is the innermost membrane of chloroplast. It is flattened sac like membrane structure where conversion of light energy to chemical energy occurs.

Grana -Dense layered stacks of thylakoid sacs. Site for conversion of light energy to chemical energy.

Chlorophyll - It is the green pigment within the chloroplast. It absorbs energy of red and blue light, and thus appears green in colour.

MECHANISM OF PHOTOSYNTHESIS

The mechanism of photosynthesis was due to American biochemist Melvin Calvin who received Nobel prize in 1961 for discovery of chemical pathways of photosynthesis.

According to Melvin Calvin the process of photosynthesis proceeds through two stages namely light and dark reaction.

i)Light Reaction -This is the first stage when absorption of light takes place and completes within 10^{-12} to 10^{-8}secs.

ii)Dark Reaction-This is the second stage which works independent of light. The dark reactions are slow hence is the rate determining step for the whole process.

Light Reaction-

This is light dependent reaction which occurs mostly in thylakoid stacks and grana. Light energy thus captured is used for preparing energy storage molecules ATP and NADPH. These energy storage molecules are used in the second stage, i.e; dark reaction.The main outlines of light reactions are-

- light energy trapped by chlorophyll to make ATP
- Simultaneously water undergoes photolysis as follows-

$$2H_2O \xrightarrow{photolysis} 4H^+ + O_2 + 4e^-$$

- The electrons thus released reacts with the carrier molecule NADP(nicotinamide adenine dinucleotide phosphate) changing it from oxidised state NADPH+ to reduced state NADPH

  $NADP^+ + 2e^- + 2H^+ \rightarrow NADPH + H^+$

  As oxygen is liberated in light reactions, the process is also known as Oxygen Liberation Step.

- When sunlight hits the chlorophyll complex, the molecule gets excited and an electron is transferred to higher energy level of the molecule.
- Light energy entering the plant splits water as hydrogen and oxygen

  H_2O + Light energy $\rightarrow 1/2\ O_2 + 2H^+ + 2e^-$

  It has been found from isotopic studies that oxygen gas is liberated from water itself and not from CO_2.

- The electrons travel through the membrane and also pump protons through the membrane generating a proton gradient.
- This proton gradient across the thylakoid membrane is responsible for ATP synthesis.
- Light absorbed by chlorophyll drives transfer of electrons and hydrogen from water to the acceptor $NADP^+$,reducing the later to NADPH.

This storage of power also reduces carbon dioxide to more complex carbon structure of glucose the building block of life.ATP and NADPH carry energy to dark reactions

Dark Reactions - Reduction of Carbon dioxide -

Dark reaction occurs the storm of chloroplasts and are independent of light.The ATP and NADPH formed during light reaction are used to make carbohydrates by reduction of absorbed carbon dioxide.The process is known as Calvin cycle. Carbon dioxide combines with 5-carbon sugar creating a 6- carbon sugar which breaks down into glucose and fructose and finally forming sucrose.

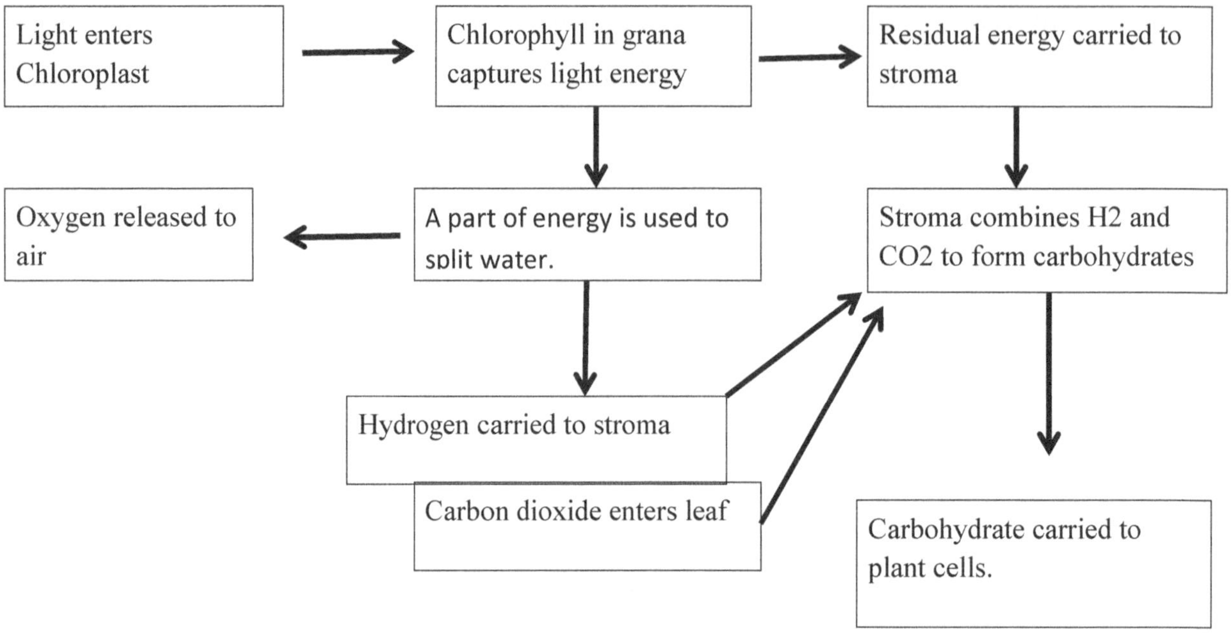

Fig: Flowchart of Photosynthesis.

The salient features of dark reactions through Calvin cycle are as follows:

i) Carboxylation -The process of fixation(i.e; addition) of molecule of CO_2 to a 5-carbon acceptor molecule ribulose bisphosphate(RuBP) in presence of plant enzyme RUBISCO is otherwise known as carboxylation. The resulting 6 carbon product splits into 2 molecule of 3 - phosphoglyceric acid(PGA) a 3- carbon product

ii) Reduction -This step consists of reduction of PGA to carbohydrate. First the PGA undergoes phosphorylation by ATP to form bisphosphate which is the reduced by NADPH to triose phosphate a three carbon carbohydrate as follows-

$$PGA + ATP + NADPH \rightarrow G3P + NADP^+ + ADP + P_I$$

Where PGA is 3-phosphoglycerate

G3P is glyceraldehyde- 3-phosphate

ADT is Adenosine diphosphate.

The $NADP^+$ and ADP so formed return to thylakoid for regeneration.

iii) Regeneration -This is the last stage in Calvin cycle that involves regeneration of RuBP the CO_2acceptor. This consist of a series of reactions starting with conversion of triode phosphate to Ribulose 5 phosphate(Ru5P) followed by phosphorylation of later RuBP.The triode phosphate can also remain in chloroplast and converted to glucose or starch.

Factors Affecting Photosynthesis Rate -

Photosynthesis process occurs at chlorophyll pigment in presence of sunlight, absorbing CO_2 from atmosphere and water from soil. The process can be affected with

i) Light intensity - Increase of light intensity increase the rate of photosynthesis.

ii) Increase in CO_2 concentration also increase rate of photosynthesis until limited by other factor.

iii) Temperature- Photosynthesis is dependent on temperature, for some cases rate of photosynthesis increase with increase of temperature for others it is the reverse.

LASERS

Laser is the acronym for Light Amplification by Stimulated Emission of Radiation. It is a device that produces a very intense beam of very highly monochromatic, coherent and well collimated light. Due to the highly monochromatic and highly collimated nature, the laser beam can be brought to a sharp focus with a lens. Due to these unique property of laser beam of light it is widely used in different fields like medicine, industry, communication ,defense, etc since its invention in 1960.

Mechanism of Production of LASER:

According to the theory of Quantum mechanics, an atom present in a lower energy(E_1) level may get excited to higher energy level(E_2) by absorbing requisite amount of energy. For example, let an atom present in lower energy level E_1 is impinged by a photon of frequency $\vartheta = (E_2 - E_1)/h$ and as a result the atom gets excited to higher energy level E_2.

This process of excitation of atom into excited state from ground state by absorbing the incident photon is called stimulated absorption.

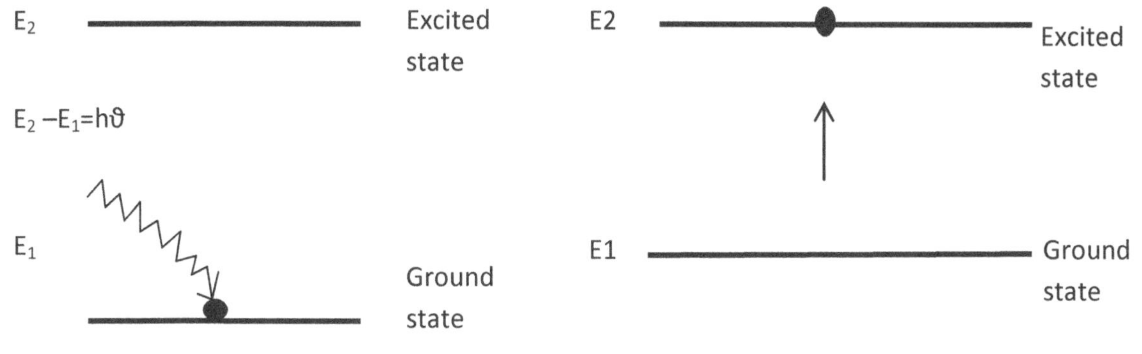

Fig: stimulated absorption.

Stimulated absorption rate depends upon the number of atoms available in the lower energy state and the energy density of photon.

An excited atom can remain at the higher state (excited state) for a limited time period of time, termed as lifetime which is of the order 10^{-8}secs.That is after a time period of 10^{-8}secs(lifetime) the atom will move to a lower energy level itself by emitting a photon during process.This de-excitation of a excited atom after its lifetime by emitting a photon energy(E_2-E_1 =h√) is known as spontaneous emission.This phenomenon was postulated by scientist Bohr.

However occasionally some atoms remain in the excited state for a longer period of the order 10^{-3}secs.This state is called metastable state.Metastable emission is very important for LASER operation.

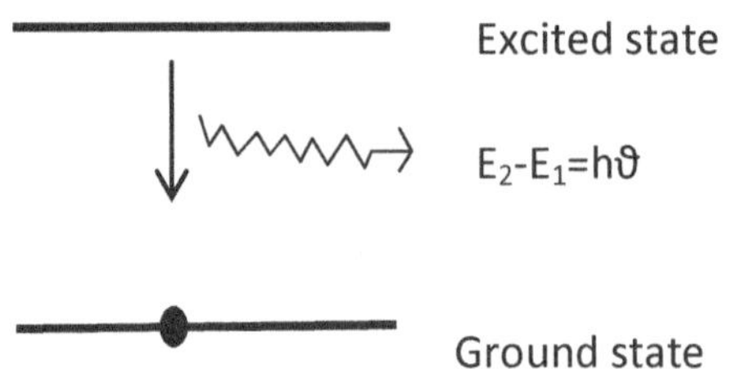

Fig: spontaneous emission.

Stimulated Emission-

Stimulated emission was postulated by Einstein. It states that if an excited atom present in the (higher) excited state E_2 is impinged by another photon of exactly same frequency $\vartheta=(E_2-E_1)/h$; then the photon will stimulate the atom to fall into the lower energy state(de-excitation) emitting another photon from the excited atom.These photons have same phase and it follows coherence.

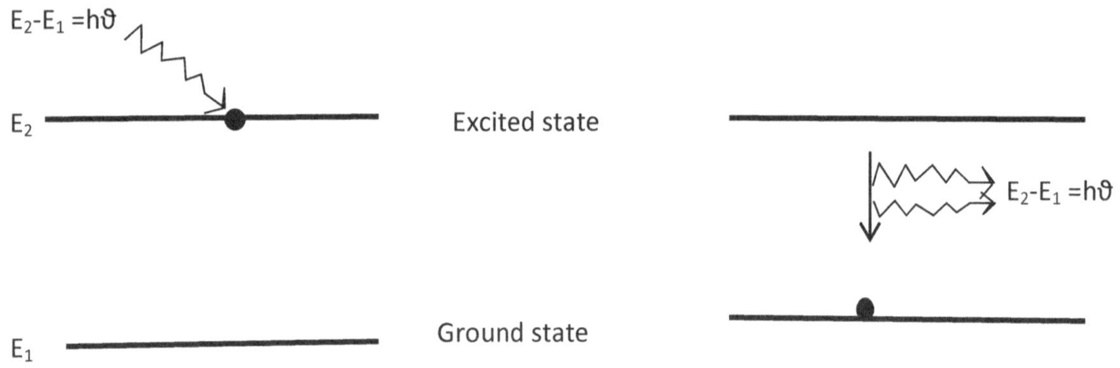

Fig: stimulated emission.

Thus incidence of radiation on an excited atom stimulates the emission of a similar radiation by the atom by transition to lower energy state.A chain reaction of similar process could be triggered by one such event to get amplification.This process is known as stimulated emission.

Stimulated emission rate $\propto$ number of atoms present in the excited state.

Stimulated emission rate $\propto$ Density of photons

Population inversion, pumping and active system -

Generally the atoms are stable in lower energy state(ground state).So the excited atoms will tend to fall to lower energy level by their own accord through spontaneous emission.

Let N_1 = Number of atoms in lower energy level

and N_2 = Number of atoms in the higher energy level.

Generally $N_1 > N_2$, but for Laser action stimulated emission should predominate over spontaneous emission.This is possible only if $N_2 > N_1$ .This can be achieve by exciting the atoms to higher energy state from lower energy state through atomic collision, by optical radiation or through other sophisticated techniques.

•In this situation when the population of atoms in higher energy level is more than population in ground(lower) energy level the phenomenon is called population inversion.

•At population inversion the population of atoms in lower energy level is called negative temperature state.

•Population inversion being a non-equilibrium situation it is achieved by artificial means as explained above and is called pumping.Or in simple words the process of raising atoms from ground state to the excited state is called pumping.

LASER Amplification Principle:-

A Laser system consists of daily three components namely:

i) Pump Source -It is also called as energy source as it provides energy to the laser system to excite atoms present at the ground state to excited state .This is achieved by passing electric discharges, flash lamps arc lamps, chemical reactions light from another laser source or through explosive devices.However the choice of pump source depends on the laser medium.

ii) Laser medium or gain medium -It is excited using pump source and causes population inversion which is followed by spontaneous emission or stimulated emersion of photons.When stimulated emission of photons is much more than spontaneous emission, laser amplification(optical gain) occurs.

iii) Optical Resonateor or Optical Cavity -An optical resonator is needed to collimate the light energy in the beam.Optical resonator usually formed by placing a pair of parallel mirrors facing each other's that light emitted along the line between the mirrors is reflected back and forth.

Light from the gain medium produced by spontaneous emissions reflected by the mirrors back into the gain medium.The reflected light through the mirrors may pass through gain medium ,many hundreds thousands of time before emission. Thereby gets amplified by stimulated emission.

Pumping Schemes-

Atoms generally consist of a number of energy levels.Among these energy levels two, three or four level will be applicable to pumping process.As two level scheme does not lead to laser action.

Three Level Lasers

A typical three level laser consist of three energy levels say E_1,E_2,E_3 , out of which E_2 is the metastable state,E_1 is the group state and E_3 is the excited state respectively.

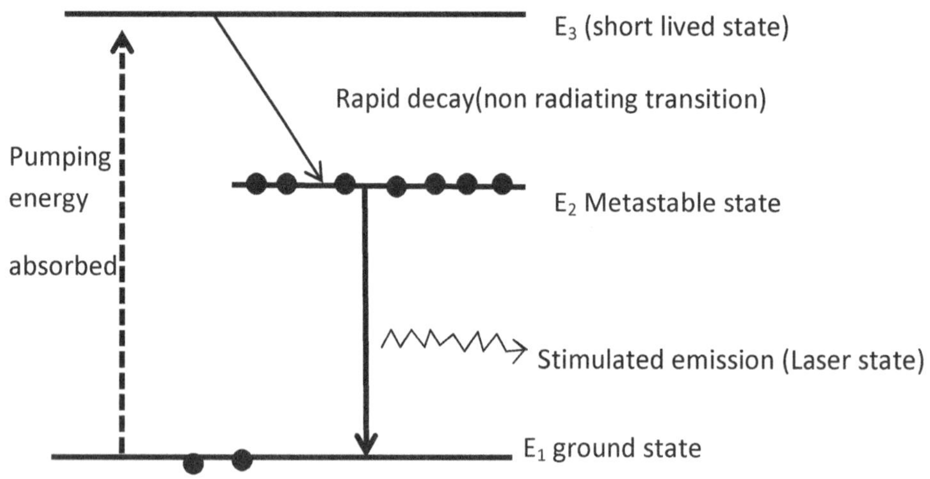

Fig: THREE level laser scheme

Intense radiation of proper frequency pumps the atomic system from ground level E_1 to excited state E_3 .The excited atoms from the excited state(short lived) rapidly decay to the metastable state E_2 .Thereby E_2 state gets heavily populated and produces the required population inversion.Now a photon of proper frequency $\vartheta=(E_2 - E_1)/h$ can cause stimulated emission from E_2 to E_1 producing Laser.If the pumping radiation is continuous, a continuous laser is obtained or it may take place in short repeated pulses to generate pulsed laser.

Four Level Lasers

As the name suggests there are four energy levels say E_1,E_2,E_3 ,E_4 .

E_1 is the ground level and E_4 is being the upper most level.E_2 is the terminal laser level above ground level E_1 while E_3 is the metastable level.The energies therefore can be represented as

$E_1, < E_2 < E_3 < E_4$

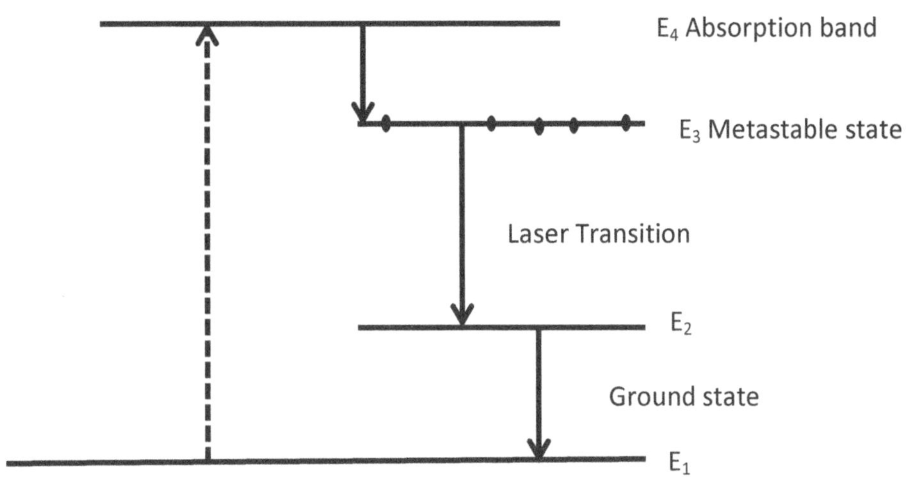

Fig: FOUR level laser system

Similar to three level laser the pumping transition elevates the atoms to short lived uppermost level E_4 .The atoms then spontaneously drops to the metastable upper levelE_3.The terminal laser level E_2 is virtually vacant and thus a population inversion between level E_3. and E_2 is quickly established.A spontaneous photon of energy $h\vartheta = E_3 - E_2$ can initiate a chain of stimulated emission culminating in lasing.This laser transition brings the atom to level E_2 .Atoms from E_2 are deexcited to $E_1$by radiative or non radiativetransition.Now the atoms present in E_1 can once again excited/pumped to level E_4 for repetition of the process.

Unlike three layer lasers the lower transition level of four layer laser is not the ground state and is virtually vacant.Thus soon as some atoms are pumped to upper laser level, population inversion is achieved.Hence less pumping energy is required for a four level lasers than three level laser.Secondly the life time of lower lesser transition level E_2 is much shorter than that of metastable state.Thus a four level laser is more efficient than three level laser.

TYPES OF LASERS

Lasers can be classified by several means namely operation mode ,element used in active medium gain of laser medium or power delivery by the laser, efficiency and so on.However a laser can be classified as in pulsed mode or in continuous wave depending on its operation.

On the basis of the material used in active medium a laser is classified as:

- Solid laser
- Liquid laser
- Gas laser and
- Semiconductor laser

Now a days with advancement of science and technology a large number of laser are used.All these lasers can be classified into two main categories like-

i) High density gain medium laser(solid and liquid laser)

ii) Low density gain medium laser (Gas laser)

Solid Lasers-

The solid laser was the first known laser.It belongs to high density gain medium type Some examples of solid state laser are as follows:

- Ruby laser
- Nd-YAG laser
- Nd-glass laser
- Turnable solid state laser
- Alexandrite laser
- Titanium-sapphire laser
- ColourCentre laser
- Fibre laser

Liquid Lasers -

The crystals used for solid state laser usually suffers from optical strain, internal defect etc hence ultimately cracks down or shatters. To overcome the disadvantage homogenous liquid solution of certain dyes in water or alcohol as solvent is used as active media. Dye lasers are tunable and have high gain compares to solid lasers. Dyes used for the purpose are:

- Polymethene dye
- Xanthene dye
- Coumarin dye
- Scintillator dye

Gas Lasers -

These are low density gain medium lasers. The optical quality of gas lasers are much better and do not suffer from inhomogeneity. Some examples of gas laser are:

- He-Ne laser
- Argon ion laser
- He-Cd laser
- Krypton ion laser
- Copper vapour laser
- Gold vapour laser

- CO_2 laser
- HF laser
- Nitrogen laser etc.

Depending upon the active media used gas laser can be categorised as atomic, ionic and molecular laser.

Semiconductor Lasers -It is specially designed p-n junction diode which emits coherent radiation of the order gigahertz wavelength and can modulate optical output. Two types of semiconductor lasers are formed ,namely- homojunction and heterojunction lasers respectively. Semiconductor lasers working at room temperature in continuous wave mode are produced by1970.

EXAMPLES OF LASERS

1. **Ruby Lasers -**

Ruby lasers is the first working laser made by Theodore H,TedMaiman in the year 1960.It is a solid state laser that uses ruby crystals as its gain medium.A synthetic ruby crystal is prepared by dipping Al^{+3} ions of Al_2O_3 crystals with Cr^{+3} ions at a concentration of about 0.05% by weight.Cr^{+3} ions have a set of three energy levels suitable for the laser action hence form the active centre while the aluminium and oxygen atoms are inert.The figure below represents the schematic diagram of a ruby laser.The xenon flash lamp excites the Cr^{+3} to the higher energy level from where it falls to the metastable state(having a lifetime of about 0.003 sec) by losing its energy to other ions of the crystal.Photon from the spontaneous decay of Cr^{+3} ions are reflected a number of times between the mirrors ends of the ruby rod thereby stimulating other Cr^{+3} ions to radiate.After a few microseconds a large pulse of monochrome coherent red-light is generated at the partly transparent end of the ruby rod.The rod length is fixed precisely at an integral number of half wavelength.So radiation trapped in it forms a standing wave.Since the induced emission are stimulated by standing wave, their waves are all in step with it.

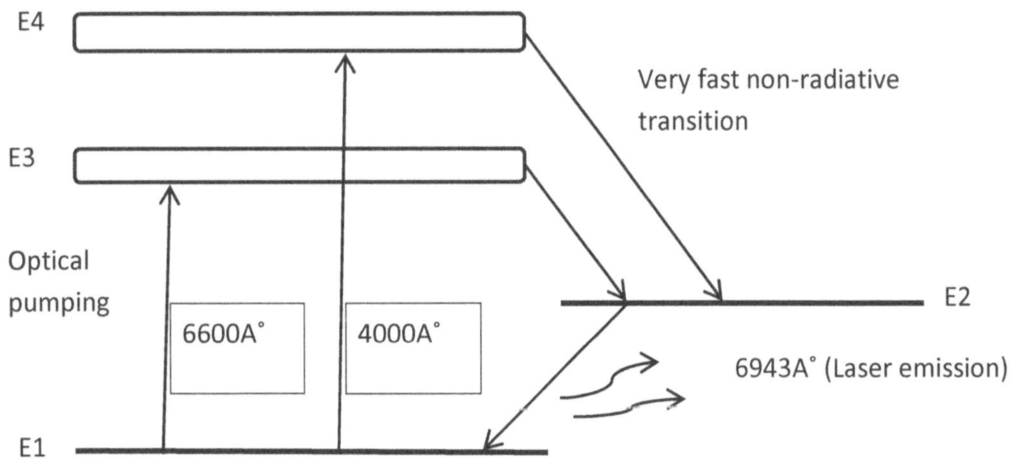

Fig: Ruby Laser

2.Nd:YAG Laser -

Nd:YAG is an acronym for neodymium-doped aluminium garnet(Nd: $Y_3Al_5O_{12}$) a compound uses a gain medium for solid state lasers.The doping concentration of Nd in Yag is of the order 0.725% by weight i.e;1.4x 10^{26} atoms of Nd per cubic metre.Neodymium is a four level system, hence need lower pump energies than ruby laser.

The Nd:Yag laser is pumped using an intense flash of white light from a xenon flask lamp.It excites Nd^{+3} ions from ground state to multiple energy state E_4 .The excited Nd^{+3} quickly decays to metastable upper laser level E_3 releasing excess energy to crystal lattice.The lower laser evel E_2 is located 0.25 ev above ground state E_1,it can not be populated at normal operating temperature.Thus helping in achieving population in version between E_3 and E_2state.In the E_3 state Nd^{+3} ions are stimulated to emit on the main 1.064μm laser transition and drop to the lower laser level E_2.From E_2 Nd^{+3} ion de-excites to the ground state E_1,by transferring energy to the crystal.The laser output is in the form of pulses occurring at a very high rate.The overall efficiency is 0.1 to 1% range.

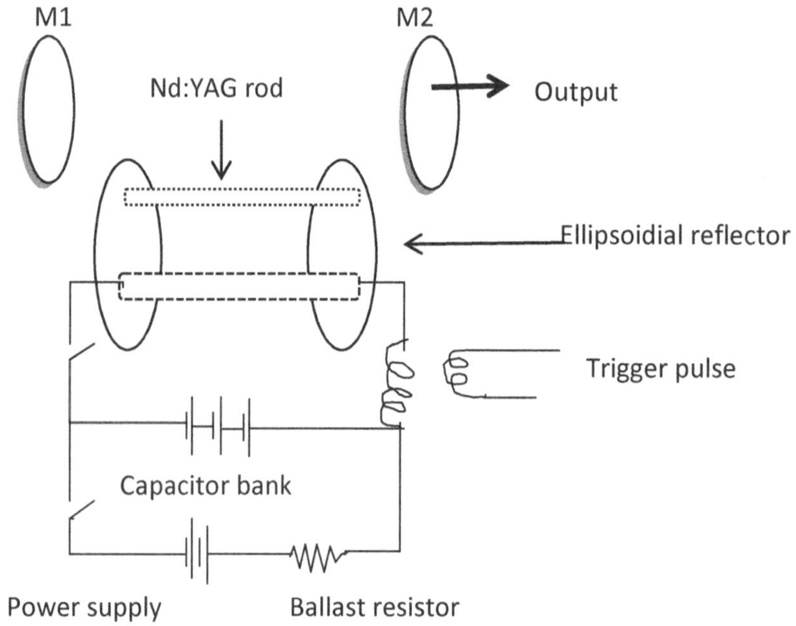

Fig Nd:YAG laser arrangement

Nd:YAG lasers are used for manufacturing of cutting and welding steel and super alloys.It is also used in making subsurface marking in transparent materials like glass or acrylic glass.

3. Helium - Neon Laser (Gas Laser) -

He-Ne atomic laser belongs to gas laser type.Atomic laser are characterized by mixing active gas with other gases which increases excitation efficiency of gases.He-Ne laser is a four level system.In He-Ne laser Ne atoms are active centres, while He atoms helps efficient excitation of Ne atoms.

A He-Ne laserconsits of a glass discharge tube of used diameter 0.5cm and length 1.5cm filled with mixture of He and Ne gases in the ratio of 10:1.This is the gain medium. The pump or energy source of the laser is provided by a high voltage electric discharge of about 10Kv.The tube is hermetically sealed with glass windows.One end of the tube is fitted with a perfect reflector while other with a partial reflector.

The passage of high voltage electric discharge ionizes the gas mixture into ions and electrons.He atoms being lighter gets easily excited to energy level E_2, while Ne atoms also gets excited to energy level E_2'.E_2 and E_2' are almost identical in energy and are metastable states.The excited He atom can not return to the ground state by spontaneous emission, but the excited He atom can return to the ground state by transferring their energy to Ne atom through collision.Consequently a large population is build up in the metastable state of Ne.Photons of frequency $\vartheta = (E_2' - E_1')/h$ then stimulate the transition to E_2' with a radiation of characteristic wavelength of 6330A° in the red region of visible spectrum.The He-Ne laser beam is highly collimated coherent and monochromatic.

He-Ne laser is widely used in laboratories in interferometry, laser printing, bar code reading etc.In supermarkets the He-Ne laser is widely used to read digitally encoded bar codes located on products.

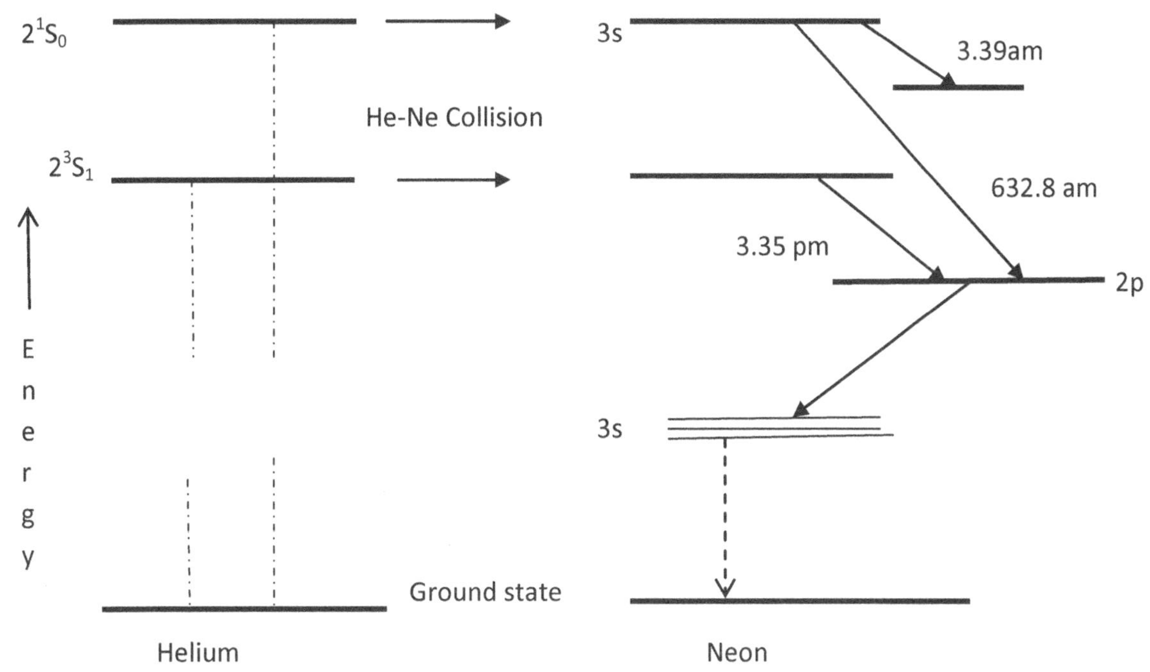

Fig: Energy Level diagram of a He-Ne laser

TABLE: Comparative study between Solid,Liquid and Gas Laser.

Sl.No.	Solid Laser	Liquid Laser	Gas Laser
1	Active medium is solid	Active medium is liquid or a liquid dye	Active medium is gas at low pressure
2	Solid laser are crystalline or amorphous solids	It is solution of dye in organic or inorganic solvents.	It is either ionic, atomic or molecular state.
3	Pumping is done usually by optical method.	Liquid laser is usually optically excited by another laser source.	It is excited by passing electrical discharge.
4	Emits a narrow range of wavelength	Emits a broad range of wavelength	Emits a narrow range of wavelength
5	Due to imperfections and defects found in crystals the output may be inhomogeneous	Output is optically homogeneous	Output is highly homogeneous.
6	Examples:Rubylaser,Nd:YAG ,Nd-glass fibre laser, etc.	Example:Pulse dye laser,tunable pulsed and continuous wave dye laser etc.	Example:HeNe, ,Argonion,Krypton ion,Co_2 laser ,etc.

APLLICATIONS OF LASERS -

Lasers are well collimated, coherent, highly monochromatic and extremely powerful beams of electromagnetic radiations, hence finds wide applications in almost every fields.The following is a list of its few applications:

- Manufacturing industry
- Metrology
- Medical
- Data storage
- Communication
- Displays
- Spectroscopy
- Microscopy
- Energy Technology
- Military

- Science and Technology.

•Medical -Lasers are a boon in medical field. Lasers are used to fuse detached retinas, Lasik treatment or corrective eye powers, treatment of cancers, burns, calcareous growth, bloodless surgery or micro surgery as highly focused laser beams can cut and coagulate simultaneously. For destruction and removal of kidney and gall bladder stones.Also effective in angioplasty and bypass surgery of heart. Endoscopy is also based on laser techniques(lithotripsy).

•Industry - It is used in cutting, boring holes in metals, materials and gem stones. Cutting and welding of metals. Lasers can cut 50 layers of cloth at time. Laser are also used to generate holograms.Lasers are also used for heat treatment of the engine cranck shaft and cylinder wads to hardenthem.Laser is used for selective tempering of specific areas also.

•Communication - Due to narrow bandwidth lasers are used in microwave communication between earth moon and other satellites.Lasers can also transmit data through fibre optic cables with a great speed and efficiency.By using laser the distance between earth to moon has been measured.

•Data Storage -The storage capacity of computers are highly improved by incorporation of laser technique.Laser printers also employed for its high quality transformation of digital data on paper.Thus lasers are used to store memory in the optical memory unit in the computers and used to read and write the beams in laser printers.Lasers are also used to read the barcodes of the products.

•Chemical Application - Laser technology is used to initiate, hasten or terminate chemical reactions.The elucidation of molecular structure is also done with laser beam.Raman spectroscopy is also a boon of laser study.Laser beams are used to carry out nuclear fusion reaction too.

•Defence Application -Laser guided missiles are used in aircrafts for pinpoint destructions of enemies missiles and artifacts.

MASERS

Maser is the acronym for Microwave Amplification by Stimulated Emission of Radiations.Unlike lasers, the masers produce coherent electromagnetic waves of microwave or radiowave frequency through stimulated emissions.

The phenomenon of masers are very common in interstellar space due to availability of required conditions like very low density and very low pressure.For example the He emission from these natural occurring masers in distinctive due to its brightness.

For the first time a maser in interstellar space was found by Harold Weaver et al of Berkeley while studying radio wave emission by -OH group in Orion Nebula(nearest galaxy).They found a very bright emission from one part of the nebula which was later confirmed to be a natural maser from the -OH group.The molecule(bearing -OH group) were being pumped by infrared light from nearly gas clouds, which were collapsing to form stars.Following this a large number of interstellar -OH maser were discovered.

In 1952 theoretical principles governing the operation of masers were first described by Joshep Weber of University of Maryland which was published in 1953.

Independently Charles Hard Townes,James P Gordon and Herbert J.Zeiger of Columbia University developed the first maser oscillator in 1953.The device used stimulated emissions in a stream of energised ammonia molecules to produce amplification of microwaves at a frequency of about 24 gigahertz.Subsequently Charles Hard Townes and Arthur L Schawlow described the principle of optical maser or laser of which T.H.Maiman built the working model in 1960.

The research of C.H.Townes,Basov and Prokhorov in field of stimulated emission is recognised by the award of Nobel Prize in Physics in the year 1964.

Applications -

- Masers are used in atomic clocks as time keeping device.
- Maser are used in an extremely low noise microwave amplifiers in radio telescopes and deep space spacecraft communication with ground stations.

Differences between Lasers and Masers

S.No,	LASER	MASER
1	The acronym is Light Amplification by Stimulated Emission of Radiations.	The acronym is Microwave Amplification by Stimulated Emission of Radiation.
2	Laser produce well collimated, coherent electromagnetic radiations in the red light regions of visible spectrum.	Masers produce well collimated, coherent electromagnetic waves of microwave or radiowavefrequency by stimulated emissions.
3	The photon from a laser comes in the form of visible light.	The photon from a maser comes in the form of a microwave or radio wave.

Questions

1.What is photochemistry?

2. What do you know about Thermal or dark reaction?

3. Differentiate between Thermal or photochemical reactions, giving emphasis on their free energy change.

4. Define Lambert's Law.

5. What is Beer-Lambert Law?

6. State Stark-Einstein law of photochemical equivalence.

7. What is Bathochromic shift?

8. Write a note on Grothus-Draper's Law?

9. Write a short note on Quantum yield or quantum efficiency?

10. Write a short note on third law of photochemistry?

11. Name the fundamental laws of photochemistry.

12. What is fluorescence? How it is different from phosphorescence?
13. Define fluorescence?How it is related to delayed fluorescence?
14. Define phosphorescence,mentioning names of two phosphorescent substances.
15. What is photosensitization?Explain with example.
16. Name a photochemical reaction that takes place in nature.
17. Write a short note on mechanism of photosynthesis.
18. Name the factors that regulate photosynthesis.
19. Write short notes on -
    a. Photosensitization
    b. Molecular excitation
    c. Phosphorescence
    d. Intersystem crossing
20. What is LASER?How does it differ from an ordinary source of light?
21. Describe the terms -
    i) Population Inversion
    ii) Pumping
    iii) Gain of Laser
22. Differentiate between spontaneous emission and stimulated emission.
23. Write note on comparison between a three level laser with a four level laser.
24. What is population inversion?Why it is necessary?
25. Mention a few engineering application of LASER?
26. What is coherent absorption?
27. Mention few medical application of LASER?
28. Name the active material in ruby laser.Draw a neat labeled diagram of ruby laser.
29. Write the role of Ne in He-Ne laser.
30. Explain the working of He-Ne laser with suitable diagram.
31. Discuss the construction and working of He-Ne laser.
32. Explain the appropiate energy level diagram of how stimulated emissions results from electron impact pumping in He-Ne laser?
33. Write short note on Nd:YAG laser.

34. Write a comparative study of solid, liquid and gas laser.

35. Why a gas laser emit light which is more unidirectional and monochromatic as compared to a solid state laser?[Hint:Due to absence of effects like crystalline imperfection, thermal distortion and scattering which is present in solid state laser]

36. What are the various applications of lasers?

37. Write a short note on history of MASERS development?

38. What is MASER? Explain.

39. Describe the role of light reaction and dark reaction in mechanism of photosynthesis.

Name Reactions

Chapter at a Glance:

Name reactions of

Aldol condensation,

Canizzaro reaction

Beckmann rearrangement,

Hoffmann rearrangement and

Diels Alder reaction.

ALDOL CONDENSATION

Aldehyde and ketones having at least one $\alpha-$Hydrogen atoms undergo self condensation in presence of a base to form β-hydroxy aldehyde respectively, known as Aldol condensation.

$$2CH_3CHO \xrightarrow{OH^-} CH_3 - CH(OH) - CH_2CHO$$

Acelatdehyde β-hydroxy butyraldehyde (Aldol)

The aldol so formed looses a molecule of water on heating to yield an unsaturated aldehyde or ketone.

$$CH_3CH(OH)CH_2CHO \xrightarrow{Heat} CH_3CH = CHCHO + H_2O$$

Aldol Crotonaldehyde

If an acid is used as a catalyst instead of alkali the unsaturated compound is directly formed due to dehydrating action of the acid as follows:

$$2CH_3COCH_3 \xrightarrow[-H_2O]{H^+} (CH_3)_2C = CHCOCH_3$$

or, $2CH_3CHO \xrightarrow[-H_2O]{H^+} CH_2CH = CH.CHO$

Aldol condensation can occur between

a) Two aldehydes (identical or different) containing α - hydrogen atoms.

$$CH_3CHO + CH_3CHO \xrightarrow{OH^-} H_3C-\underset{}{\overset{\overset{OH}{|}}{C}H}-CH_2-CHO$$

β – hydroxy butaraldehyde

b) Two ketones (identical or different)

$$CH_3COCH_3 + CH_3CHO \xrightarrow{OH^-} (H_3C)_2 - \overset{\overset{OH}{|}}{C} - CH_3COCH_3$$

4-Hydroxy-4-Methyl-2-pentanone

c) An aldehyde and a ketone both containing α - hydrogen atoms.

$$CH_3COCH_3 + CH_3CHOCH_3 \xrightarrow{OH^-} H_3C - \overset{\overset{OH}{|}}{C}H - CH_2 - CO - CH_3$$

4-Hydroxy-2-pentanone

d) and condensation between an aldehyde or ketone having no α - hydrogen atom with ketone or aldehyde having α - hydrogen atom

$$H - CHO + CH_3 - CHO \xrightarrow{OH^-} \underset{\underset{OH}{|}}{C}H_2 - CH_2 - CHO$$

β – hydroxy pionaldehyde

MECHANISM :

The base catalysed aldol condensation takes place in three steps.

NaOH furnishes OH^- ions

$$NaOH \rightleftharpoons Na^+ + OH^-$$

1) The first step consists of abstraction of an acidic hydrogen atom from the α –position of the carbonyl group to generate a carbanion (1). This is the rate determining step of the reaction.

$$H\bar{O} + H - CH_2 - \overset{\overset{O}{\|}}{C} - H \underset{-H_2O}{\Leftrightarrow} \bar{C}H_2 - \overset{\overset{O}{\|}}{C} - H$$

Carbanion (1)

2) The carbanion then attacks the carbonyl group of other molecule producing electromeric shift of π electrons and adds to the electron deficient carbon of the carbonyl group forming anion (II)

$$R-\underset{\substack{|\\ H}}{\overset{\substack{O\\ \|}}{C}}+C^{-}H_2-CHO \longrightarrow R-\underset{\substack{|\\ H}}{\overset{\substack{O^{-}\\ |}}{C}}-CH_2CHO$$

(Anion II)

The slower rate of attack of the carbanion the carbonyl carbon atom due to +I effect of methyl group.

3. The anion (II) takes up a proton from water and forms an aldol

$$R-\underset{\substack{|\\ H}}{\overset{\substack{O^{-}\\ |}}{C}}-CH_2-CHO+\quad H-O-H \longrightarrow R-C-\underset{\substack{|\\ H}}{\overset{\substack{OH\\ |}}{C}}H_2-CHO+OH^{-}$$

H (ALDOL)

The aldol so formed on dehydrating (on heating or in presence of acid) eliminates a molecule of water to produce α,β unsaturated compound.

$$R-\underset{\substack{|\\ OH}}{C}H-\underset{\substack{|\\ H}}{C}H-CHO \xrightarrow[-H_2O]{H^{+}\Delta} R-CH=CH-CHO$$

α,β-unsaturated aldehyde

As the aldol still contains a free carbonyl group it can further add a molecule of aldehyde or ketone as follows:

$$R-\overset{\substack{OH\\ |}}{C}H-CH_2CHO+CH_3-CHO \xrightarrow{OH^{-}} R-\overset{\substack{OH\\ |}}{C}H-CH_2-\overset{\substack{OH\\ |}}{C}H-CH_2CHO$$

$$\Big\downarrow -2H_2O$$

$$R-CH=CH-CHO+CH_3CHO \xrightarrow[-H_2O]{OH^{-}} R-CH=CH-CH-CH-CHO$$

The above mechanism is supported by the fact that the reaction is a second order reaction, first order with respect to both the acetaldehyde and hydroxide ion. If the initial carbanion formation was fast and reversible, the reaction would have been second order in acetaldehyde.

CANNIZARO REACTION

Aldehydes which do not contain any α-hydrogen atom undergoes Cannizaro's reaction. In this reaction two molecules of an aldehyde lacking α-hydrogen atom(Ex. HCHO, C_6H_5CHO, CCl_3CHO etc) undergoes disproportionation yielding corresponding alcohol and acid in presence of concentrated alkali.

$$2H-\overset{\overset{O}{\|}}{C}-H \xrightarrow[\text{Room Temp}]{\text{Conc. NaOH}} H_3COH + H-\overset{\overset{O}{\|}}{C}-ONa$$

Methanal Methanol Sod. Formate

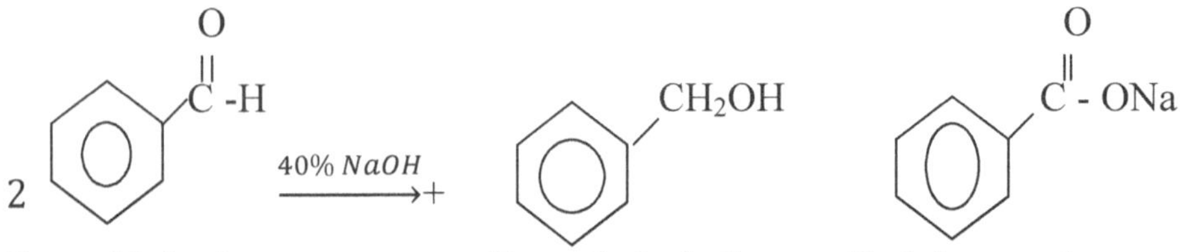

Benzaldehyde **Benzyl alcohol** **Sod. benzoate**

MECHANISM

From experimental observation followed by kinetic studies with isotopically labelled compounds the following general mechanism had been suggested for cannizaros reaction taking benzaldehyde as an example

The first step consists

1. Addition of the nucleophile to the aldehyde molecule

$$Ar-\overset{\overset{H}{|}}{C}=O+OH^- \longleftrightarrow Ar-\underset{\underset{OH}{|}}{\overset{\overset{H}{|}}{C}}-O^-$$

Forming anion Anion

2. Second step comprises of transfer of hydride ion from anion to a second molecule of the aldehyde.

$$Ar-\overset{\overset{H}{|}}{C}=O+Ar-\underset{\underset{H}{|}}{\overset{\overset{OH}{|}}{C}}-O^- \longrightarrow Ar-\underset{\underset{H}{|}}{\overset{\overset{H}{|}}{C}}-O+Ar-\overset{\overset{OH}{|}}{C}=O$$

Alkoxide ion Acid

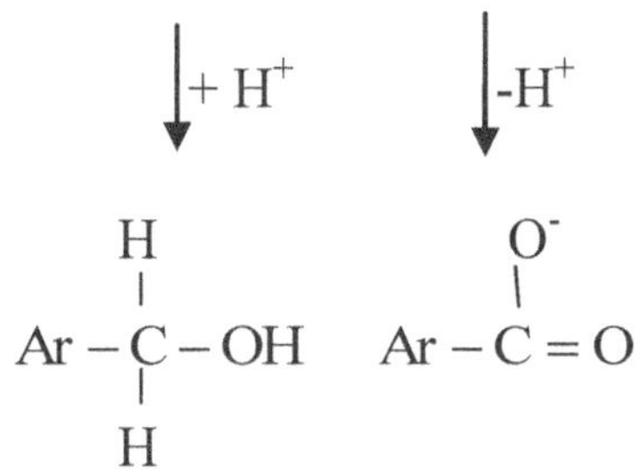

This is the rate determining step of the reaction. The alkoxide ion and the acid ion so obtained, then exchange proton to yield more stable pair namely alcohol and acid anion.

Trimethyl acetaldehyde responds to Cannizzaro reaction as follows:

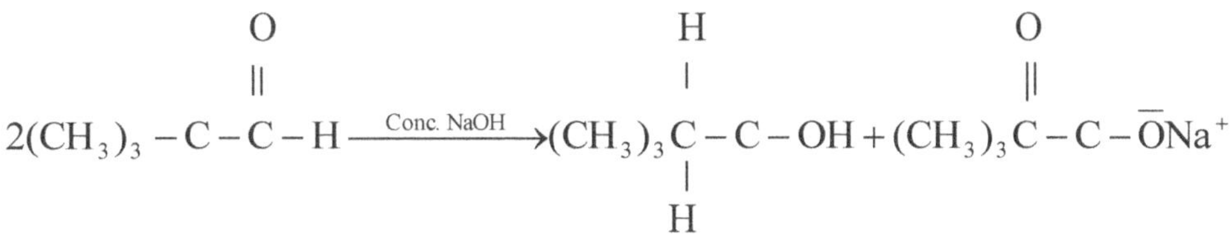

Intramolecular Cannizzaro reaction – When cannizzaro's reaction takes place within the same molecule. Half part of the aldehyde is oxidized and other half is reduced as follows:

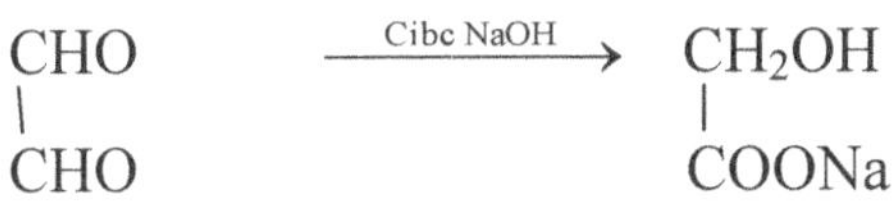

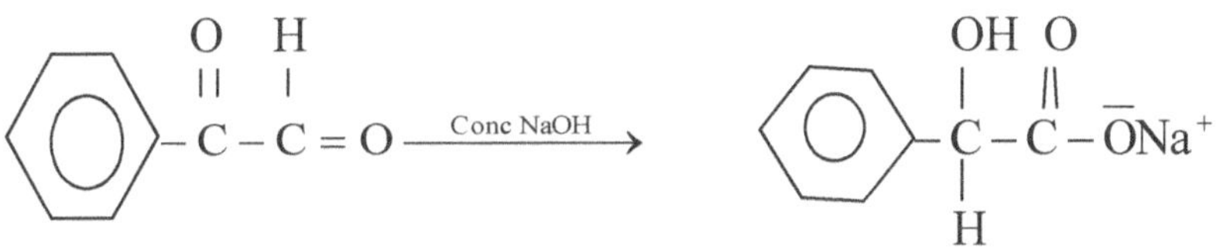

This process is named as intermolecular Cannizzaro reaction

Crossed Cannizzaro Reaction

A mixture of two aldehydes allowed to undergo Cannizzaro reaction following products are obtained and the process is termed as Crossed Cannizzaro reaction.

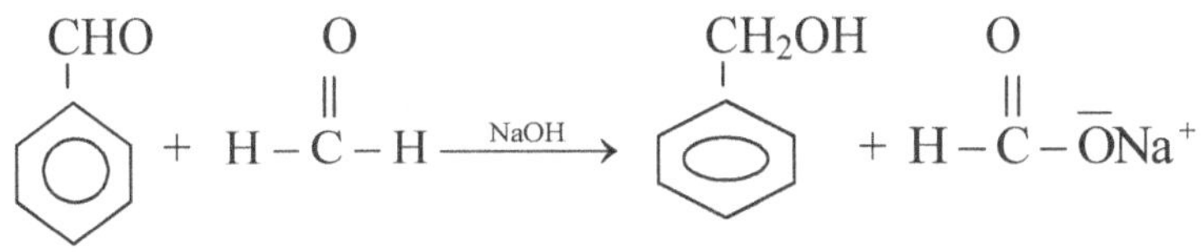

BECKMAN REARRANGEMENT

Aromatic ketoximes when treated with reagents like phosphorous pentachloride, sulphuric acid, or polyphosphoric acid undergo rearrangement to produce substituted amides. The phenomenon is named as Beckmann rearangement after the name of the discoverer.

$$Ar_2C = NOH \longrightarrow ArCONHAr$$

Aromatic Ketoxime Substituted amide

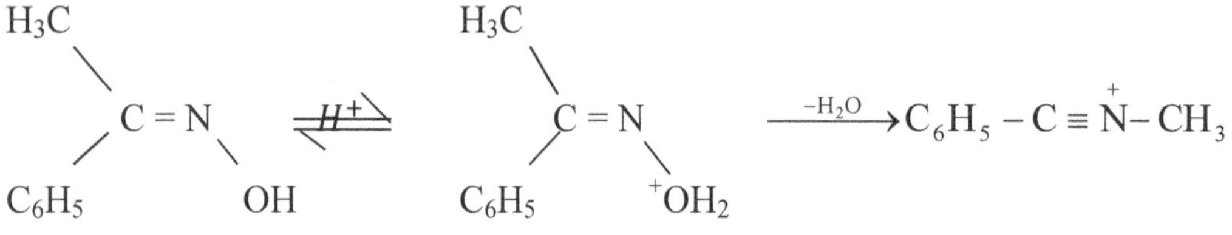

Acetophenone Oxime

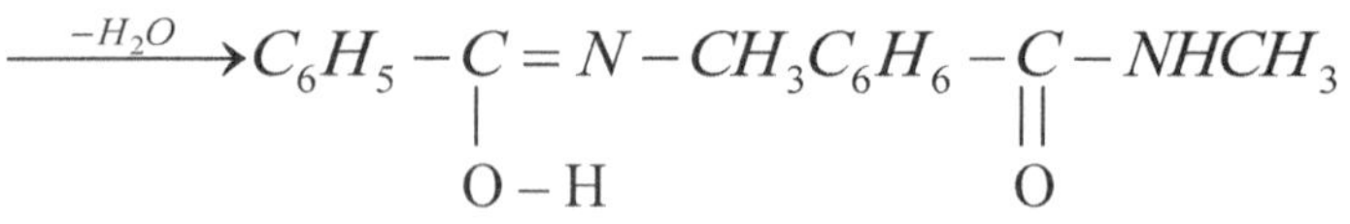

Acetanilide

The Beckmann rearrangement is stereospecific in nature and always intramolecular. The migrating group is also always trans to the hydroxyl group. Hence by identifying the amide and using the stereospecificity of the rearrangement , configuration of the ketoxime can be determined.

However the mechanism of the Beckmann rearranging is controversial as the rearrangement can occur under a wide variety of catalyst and even in absence of a catalyst.

For example Benzophenone oxime undergoes rearrangement in presence of benzene sulphonyl chloride and an alkali and benzophenone oxime picrate rearranges only on heating.

Like other heterocyclic reactions the rearrangement is accelerated by heat and is faster in polar solvents to non-polar one.

The oxime does not rearrange itself but rearrangement takes place in presence of a reagent. It is believed that some intermediate is formed between the oxime and the reagent which undergoes rearrangement.

For an aryl migrating group containing electron withdrawing substituent in para position the rate of rearrangement is slower as in rearrangement of picrate derivative of benzophenone oxime. While with electron releasing substituents in aryl group the rearrangement is slower due to involvement of phenium ion intermediate.

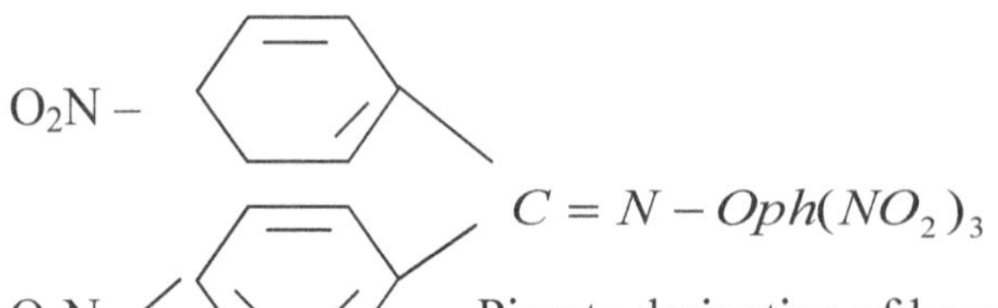

Picrate derivative of benzophenone oxime

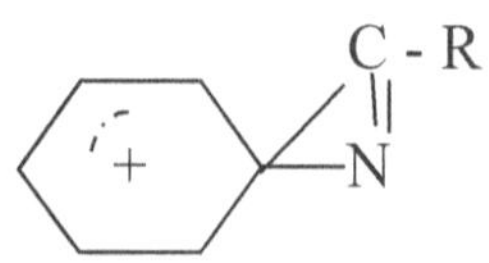

Phenium ion

The rate rearrangement also depends on the nature of the leaving group

As rate of anion elimination from the oxime depends directly on the strength of the acid.

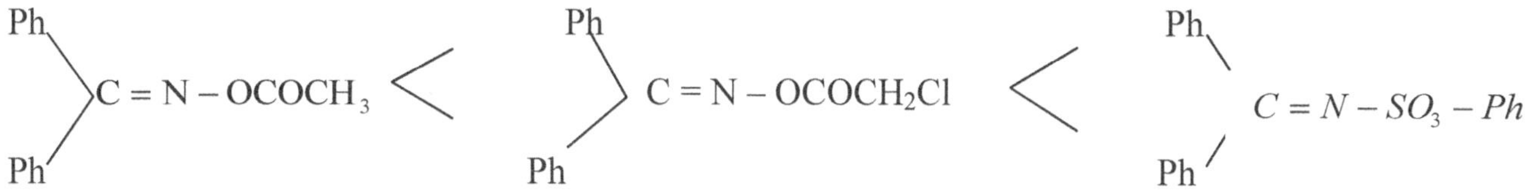

Beckmann rearrangement is widely used for the synthesis of ε-caprolactem (from cyclohexanone oxime) the precursor for production of the condensation polymer Nylon-6.

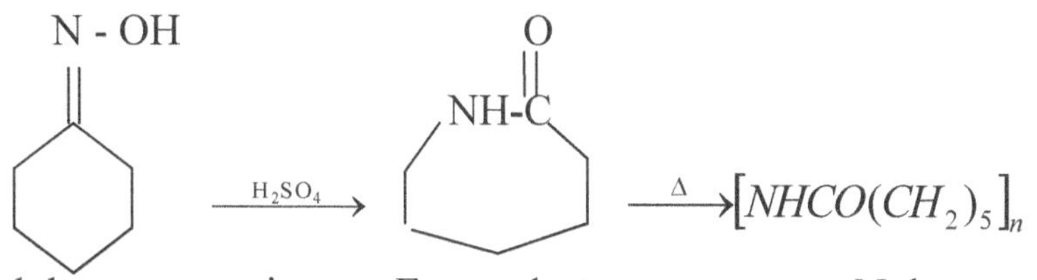

Cyclohexanone oxime   E-caprolactum   Nylon

HOFFMANN REARRANGEMENT

In this rearrangement an unsubstituted amide when treated with sodium hyprobromite (or bromine in alkali) produces N-bromo amide which in presence of a base gives its conjugate base. The conjugate base so formed undergoes rearrangement to form isocyanate.

$$RCONH_2 + NaOBr \longrightarrow RCONHBr$$

Mechanism

$$RCONHBr \underset{}{\overset{OH^-}{\rightleftharpoons}} \left[R-\underset{\underset{O}{\|}}{C}-\bar{N}-Br \longleftrightarrow R-\underset{\underset{O^-}{|}}{C}=N-Br \right]$$

$$\xrightarrow{-Br^-} O-C=N=R$$

Isocyanate

However the isocyanate formed is rarely isolated, but depending on the reaction condition aqueous or alcoholic solution it yields an amine or urethane respectively.

$$R-N=C=O \xrightarrow{H_2O} R-NH_2 + CO_2$$

$$R-N=C=O \xrightarrow{R'OH} RNHCOOR'$$

Urethane

Mechanism

The N-bromoamide loses a proton to the base and is acidic due to presence of two electron withdrawing groups (acyl and halo) on the nitrogen. This step is followed by removal of Br^-

(bromide) to yield the nitrene (isocyanate). This mechanism is confirmed by isolation of N-Haloamide as an intermediate.

Similarly $\beta-$amido pyridine gives $\beta-$aminopyridine and phthalimide gives anthranilic acid.

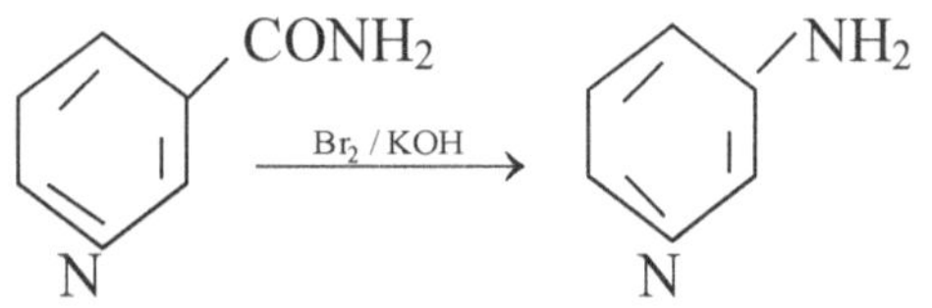

β -amido pyridine β -amino pyridine

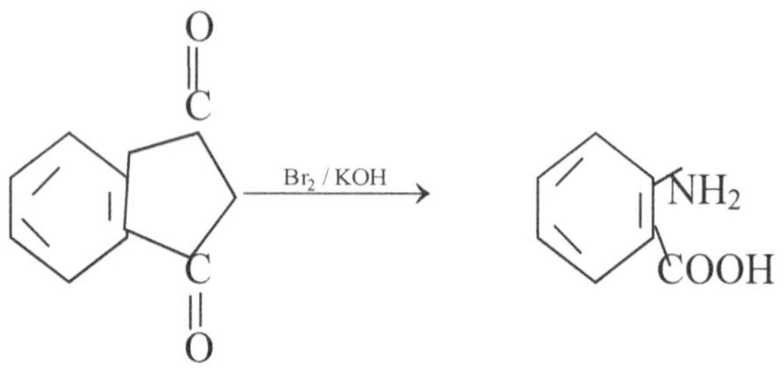

Phthalimide Anthranilic acid

DIELS-ALDER REACTION

This reaction involves interaction of a conjugated diene with an unsaturated molecule (dienophile) to form a six membered ring compound called an "adduct" without elimination of any molecule.

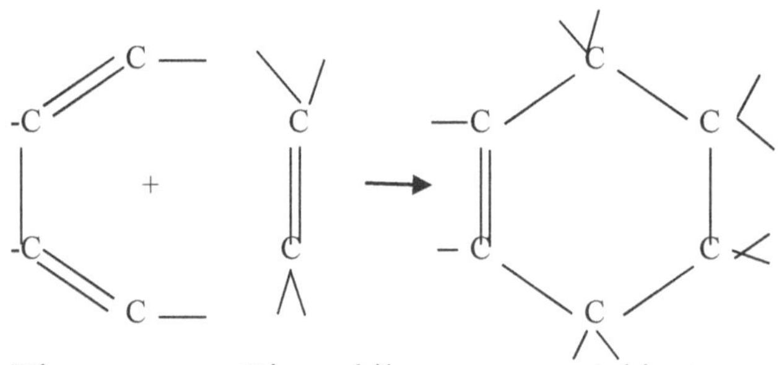

Diene Dienophile Adduct

The reaction involves 1,4 addition of a dienophile to a conjugated diene to produce a cyclic compound. During this cyclo addition two π bonds are changed/converted to two σ bonds. As this cycloaddition involves combination of a 4π system (i.e. 1, 3-diene) with a 2π system (dienophile) is also known known as (4+2) cyclo addition.

MECHANISM

The mechanism of Diels Alder reaction is yet to be established. It can follow a heterocyclic (ionic) mechanism and also a free radical pathway as follows:

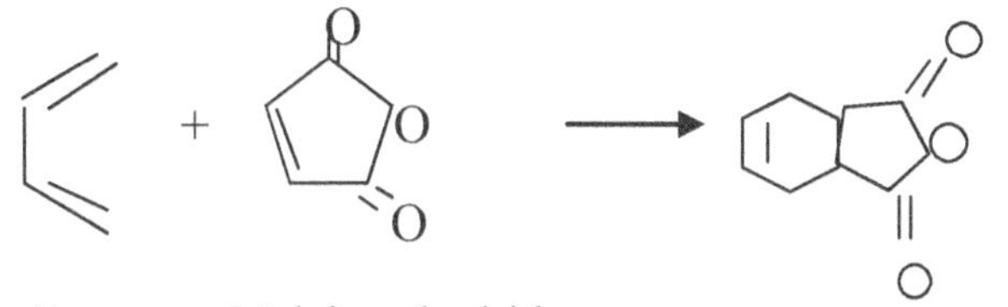

Butadiene　　Maleic anhydride　　Tetrahydrophthalic anhydride

IONIC MECHANISM

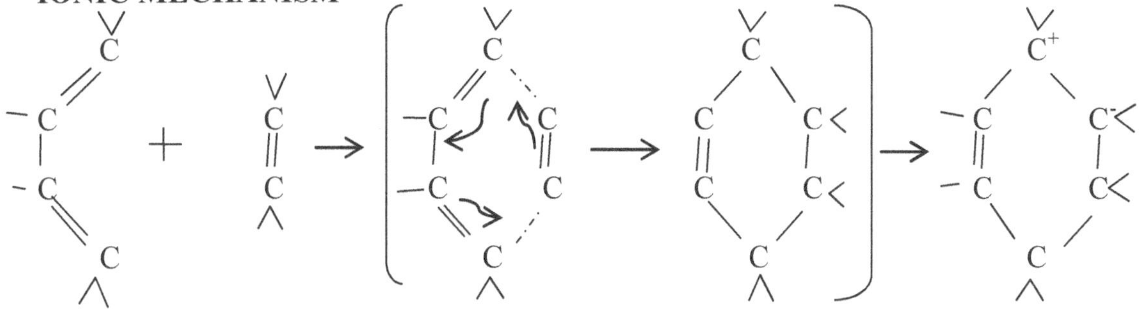

Free Radical Mechanism

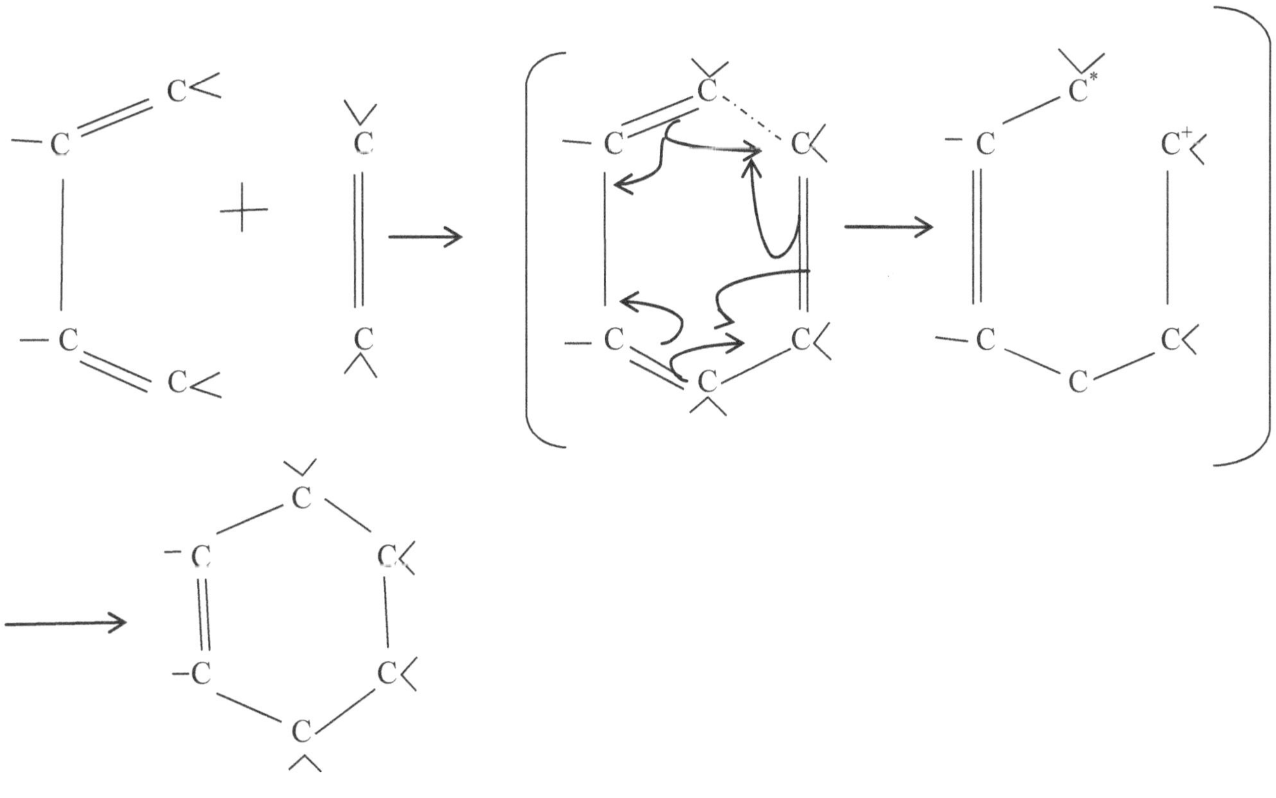

However in both the cases a redistribution of electron density takes place resulting in new.

- Formation of 2 new single bonds.
- Conversion of 3 double bonds to single bonds
- And conversion one single bond to double bond

In both the cases the reaction follows a second order mechanism (first order with respect to diene and first order with respect to dienophile. The cyclo addition is less sensitive to polarity of the solvent but yields cis-addition products. However the rate of the reaction is accelerated by presence of any electron withdrawing group on the dienophile and electron donating group on the diene and the vice versa.

Diels Alder reaction is widely used for its retention of stereospecificity property.

- It is used for synthesis of an alkaloid called reserpine used for hypertension drugs.
- Diels Alder reaction helps in stereochemical fixation of configuration in the synthesis of chloresterol.

Index